Connected Healthcare for the Citizen

Health Industrialization Set

coordinated by
Bruno Salgues

Connected Healthcare for the Citizen

Edited by

Robert Picard

First published 2018 in Great Britain and the United States by ISTE Press Ltd and Elsevier Ltd

ISTE Press Ltd
27–37 St George's Road
London SW19 4EU
UK

www.iste.co.uk

Elsevier Ltd
The Boulevard, Langford Lane
Kidlington, Oxford, OX5 1GB
UK

www.elsevier.com

British Library Cataloguing-in-Publication Data
A CIP record for this book is available from the British Library
Library of Congress Cataloging in Publication Data
A catalog record for this book is available from the Library of Congress
ISBN 978-1-78548-298-4

Printed and bound in the UK and US

Contents

Introduction

The digital world is becoming increasingly involved in the lives of the French. All of its uses are developing, both in the personal and professional worlds. According to the digital barometer[1], three quarters of the French population connect themselves to the Internet daily; one out of three workers bring their smartphone or tablet to work to use it for professional reasons. In the last few years, the smartphone has become indispensable and close to three quarters of the population own one.

The individual preoccupied by their health, the patient, the disabled person, just like medical and social workers, are concerned by this evolution. It justifies the use of the expression "citizen connected healthcare". In this expression, the reference to citizenship reminds us that this healthcare concerns the socially involved. This is the point of view presented in this book. Here, health is defined by the importance attributed to it by global healthcare organizations, for example "a state of complete physical, social and mental well-being, and not merely the absence of disease or infirmity" [OMS 46]. Finally, the notion of connection underlines the fact that the citizen's healthcare refers to our new sociotechnical reality which transforms the relationship between individuals and their healthcare system.

Introduction written by Robert PICARD.

1 Yearly reference study measuring the adoption of digital devices and services by French citizens: https://www.economie.gouv.fr/cge/barometre-numerique-edition-2017.

I.1. Context

The context which has led to the emergence of a "citizen connected healthcare" can be presented according to three different dynamics: that of healthcare linked to society, that of the economy of new technological solutions, and that of the rules and regulations of society, which created the conditions necessary for social and sanitary demands to meet technological opportunities and find satisfactory solutions to said demands. From now on, we will successively approach these topics.

I.1.1. *Context of social and public health*

Our society is increasingly involving its citizens, considering individual aspirations, allowing more autonomy and more solidarity. The digital world profoundly transforms social relationships and can allow new responses. The healthcare system is re-questioned due to the arrival of ever more efficient and targeted treatments, for economic and technological reasons, and considers research into a new approach for patients and their carers (the notion of a patient's journey through treatment). This system initiates and is affected by profound transformations.

Here emerges a new vision of the healthcare system's role, called to support the autonomy of people and their ability to freely make life choices, despite their health vulnerabilities. Notions of empowerment[2] (which refers to a patient's ability to take control of their healthcare), and of literacy (meaning a form of knowledge about the said healthcare), illustrate this point and are now in common use.

The Cap Santé report by Christian Saout [SAO 15] on supporting patients' autonomy has developed this concept and its implications. This translates into a profound evolution of the relationship between society and the healthcare system. This report has been used as a basis for launching experimental work on the conditions and evaluations concerning patient support, aiming to develop in the patient, a broader control of their health.

---

2 We will go back to this term in section 1.3.2.

I.1.2. *Techno-economic context*

The technological and digital environment is known as a transformational lever for a healthcare system undergoing a crisis, confronted with an aging population, a shortage of medical resources (or their inadequate distribution) and an increase in medical expenses.

Many international studies and reports offer statistics which underline the impressive diffusion of technologies connected to healthcare, of e-healthcare, health connected products and of connected healthcare. In 2016, 73 million healthcare devices were connected across the world. In 2020, there will be 161 million, according to a study by Grand View Research[3]. This growth will mainly consist of three trends: the rise in the average age of the worldwide population, the prevalence of diseases requiring constant monitoring in certain countries (such as diabetes) and the growing demand for quantifiable solutions.

According to the Data Bridge Market Research's 2017 market study[4], the Internet's market for medical products rose to 157 billion in 2016 with a growth of +30% per annum. Today, this market is dominated by wearables (bracelets, watches or all connected clothing), which represented 60% of the market for connected medical products in 2015.

Meanwhile, even though e-health projects are abundant, we have little information concerning future results (about observance, efficiency for the individual – quality of life, permanent ownership, etc. – or for healthcare systems – coordination between medical professionals, organization of healthcare systems). In addition, the segmentation of the offer and chain of value are complex, unstable due to the lability of demand and the fact that technological evolutions are fast. Owing to this, evidence is difficult to establish.

Consequently, it is difficult to develop value propositions, economic models, and we observe an erratic evolution in the offer in a field that has yet to structure itself.

3 https://www.grandviewresearch.com/industry-analysis/e-health-market.
4 http://databridgemarketresearch.com/reports/global-internet-of-things-iot-healthcare-market/.

The 2012 report by the Ernst and Young Society[5], which is still relevant today, was one of the first studies to shed light on the profound transformation to come concerning the medical equipment market and its products, which, despite being communicating, are becoming connected products (see section I.1.2.1).

According to this report, healthcare technologies should aim for results and behavioral evolution in the patient (Figure I.1). We would add that beyond the patient, their professional and familial circles are equally to be taken into account. These fundamental questions that are jostling the medical industry call for innovative offers. The challenge is that the medical technology market must remain viable throughout.

I.1.2.1. *A vision of connected medical technologies (according to Ernst and Young)*

The aspects of "I" for "information" and "P" for "patient" characterize the ongoing transformation[6].

"I". Processing information

It is necessary to bring to the patient – as well as the doctor and medical professionals who care for them:

– information regarding "objectives";

– monitoring over time (thanks to sensors);

– systems to assist the patient's decision-making;

– acknowledgment that the patient is "competent" (or else the medical professional has a tendency to orient treatment towards their preferences, or what they know best…).

"P". Changing the patient's lifestyle

– Healthy behaviors must be promoted, isolation must be broken when appropriate;

– incentives are effective if they provide immediate results;

---

5 "Pulse of the industry, medical technology report" 2012: http://www.ey.com/gl/en/industries/life-sciences/pulse_medical-technology-report-2012.

6 No doubt today we would also add the letter "S" for "self-determination".

– it is possible to use ICT (information and communication technology), smartphones, etc.;

– serious games are, among others, a way to influence behaviors;

– GPS can be used to ensure that the patient attends mandatory appointments, etc.

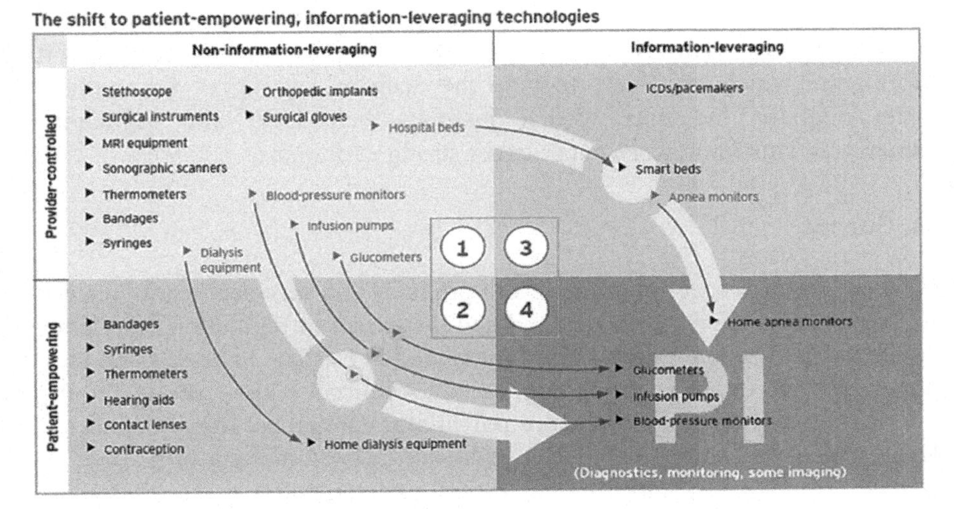

Figure I.1. *The P & I model of medical devices according to Ernst and Young's "Pulse of the industry, medical technological report"*

If the patient's new position is recognized by the industry, conversely, technologies are identified as a lever for changing the patient's behavior. In the previously cited Cap Santé report [SAO 15], technology's role is mentioned. Indeed, certain devices can reinforce one's autonomy during the daily self-management of healthcare, serving a life project; social networks can support active communities, reinforcing each member's feeling of community belonging, etc.

Responding to the expectations and needs of each ill individual is a constant imperative goal and all proposed digital solutions will necessarily be submitted. Not everyone will be ready to use them. Many are hoping for new possibilities to help with their self-care abilities. Promoting procedures aiming for clear information, a shared reflection on their possibilities and support in their methods of use, is one of the coming transformations, the one which is needed for e-health to begin to increase people's, communities' and organizations' abilities to act [HUY 16].

I.2. Regulatory context[7]

Connected objects, devices and mobile applications are increasingly used in the field of healthcare in a large way. They are carriers of personal information which are likely to help the decision-making of the medical professional or the user, which justifies their legal and regulatory framework. This framework experiences strong evolutions.

I.3. Europe

These features specific to the health products and services sector concern the heavy reliance on public demand and the extent of regulation. The latter results from "essential requirements" mediated in Europe. In healthcare, they aim to protect citizens during the use of solutions which, acting on the functioning of the human body, are potentially dangerous. Many connected devices, as soon as they are used in medical decision-making or as soon as they pose a risk to the citizen's health, fall under regulations as a "medical device" (MD).

The regulations surrounding MDs are recent (93/42/CEE of 1993) and were made mandatory in 1998. They take the form of CE marking which guarantees the respect of essential requirements. The definition of performance here is the treatment of risk and the fact of not compromising the security of the user. It has already undergone important revisions, and when reviewed by European directives, it has been subjected to different transpositions in different countries.

A new regulatory framework of MDs was published in May 2017 (2017/745): it is now a regulation, without transposition in the law of the States, which is conclusively going towards a harmonized landscape. Its application will become obligatory in all States in 2027; however, this

7 Information stemming mainly from the General Council of the Economy's report, "Attractivité de la France pour les entreprises de santé": https://www.economie.gouv.fr/cge/attractivite-france-pour-entreprises-sante.

obligation can be decided by certain States starting from 2020 at their initiative. This is also the case for France.

This framework resembles the requirements of the Union of the American FDA[8], harmonizing the evaluations of different countries by the implementation of a pool of international experts, and therefore reducing the potential inequalities of the appreciation of an agency of evaluation over another in Europe.

Among the medical devices, the technological solutions involved with the most risk – implants for example – are confronted with delays in the development of these solutions, because of their tight regulations. The development period as well as the approval of these devices can sometimes take several years, similarly to the drug development process. It must be noted that these requirements do not apply to certain medical technological solutions. However, the regulatory field of health is expanding: storing data regarding health, mobile applications as well as connected products to health that are not MDs and so on. Many of the cases explored in this book will illustrate this point.

Finally, let us remind ourselves that the regulation of the European Union no. 2016/679, known as "General Data Protection Regulation" (GDPR)[9], includes the reference text that defines the protection of personal data, including data regarding health. It reinforces and unifies the protection of data for the individual at the heart of the European Union, by simultaneously impacting development process of the products and services regarding health.

I.4. France

The French National Authority for Health (HAS) is continuously conducting work regarding the application and the evolution of the necessary regulations because of the new emerging technological solutions. The HAS has its interests particularly in mobile applications as well as products connected to health, whether they be of medical status or not, as long as they have links to the medical field.

8 Food and Drug Administration.

9 http://www.consilium.europa.eu/fr/policies/data-protection-reform/data-protection-regulation/.

The work aims for "soft" regulations to control the mobile applications and products connected to health. It took the shape of a "reference for correct practice"[10] and mobilized the national community to contribute to a labeling of the solutions on the market; it did this with the implementation of a specialized committee "Santé" within AFNOR (Association Française de Normalisation).

The aimed regulation consists of two aspects, namely the protection of the user and the demonstration of a "proven benefit":

– user protection: this entails three dimensions – the reliability of healthcare, data protection and cyber security. The evaluation criteria must be explicit, measurable and the associated methods must be identified;

– the proven benefit: this refers to use values, at the heart of the activity of design and evaluation, the Living Labs in health and autonomy – LLSA. This second point remains to be explored: the methods of participative co-design and evaluation of uses, like those practiced by Living Labs, still have effects that are not sufficiently known by the sector experts and the responsible institutional bodies in charge. The articulation between the clinical trial approach as well as these methods remains to be invented (see Chapter 8).

I.5. Purpose of this book

In this context, this book has ambitions to reflect on the challenges and the conditions of a long-term appropriation of technologies (mobile applications, connected medical devices, etc.) by our fellow citizens, aimed at the benefits to their health, and more broadly speaking, for the overall health of the population.

Such a reflection sparks an interest in all bodies: the citizen, whatever the state of their health, the medical professional who prescribes or mobilizes these new solutions, health establishments and institutions, the enterprises that produce the equipment or offer these services, etc.

10 https://www.has-sante.fr/portail/jcms/c_2681915/fr/referentiel-de-bonnes-pratiques-sur-les-applications-et-les-objets-connectes-en-sante-mobile-health-ou-mhealth. See section 8.3 for more information regarding this reference.

This is precisely the reason why the Forum LLSA has found an interest in the subject. The forum is an association that regroups the LLSA – Living Lab in health and autonomy – located in France with some other French speaking LLs, and also all types of ecosystem actors (see Box I.1). The forum regularly brings its members together around themes of reflection to share with each other, relying on concrete cases that they have experimented with (feedback).

DEFINITION.– A Living Lab is a consultation mechanism bringing together public and private actors, business professionals, funding bodies, associations and users in order to design, collectively, innovative solutions in technology, organizations and services supporting new responses for communities and society and to evaluate these solutions [PIC 17a].

The Forum LLSA – the forum of Living Labs in health and autonomy – came together in late 2013 after 2 years in development and in the extension of the report by the Conseil Général de l'Economie[11] on this theme. It started without any legal structure so as to allow a large public participation. It has equipped itself with a supporting association that includes among their founders, the Université Catholique de Lille, Institut Mines Télécom, Université de Technologie de Troyes, as well as the FEHAP – a non-profit federation of private establishments, the Institut Français de Recherche sur le Handicap. Forum LLSA members have bound themselves to a participative development of the design of new goods and services for health and autonomy, serving innovation, economic development and a democratization of health. The forum has also supplied itself with a charter[12]. See also www.forumllsa.org.

Box I.1. *The forum LLSA of Living Labs in health and autonomy*

11 Structure of the French Ministry of the Economy – www.economie.gouv.fr/cge.

12 http://www.forumllsa.org/bundles/fllsageneral/pdf/Charte_du_Forum_LLSA_24-09-13.pdf.

I.6. Method

Forum members reflecting the diversity of the health ecosystem (professionals, researchers, patients, manufacturers, etc.) met monthly in "work groups" for a year. They did this to present and share their experiences on the theme of this book, by using concrete cases, featuring connected products and the sociotechnical systems associated with them.

Each presentation of a particular case had to address a certain number of questions, and these questions were developed on the basis of a prospective reflection from the Conseil Général de l'Economie and were discussed during the first work group meeting (see Chapter 2). These cases were tackled and debated by experts of diverse disciplines. The reported experiences during the collective work and picked up on in this book illustrate how connected health is in line with significant problems of the health of the population, such as: an aging population, the increasing burden of chronic illnesses and the unequal geographical distribution of medical professionals, in this context of a "connected society".

It became more a case of presenting what caused the problems rather than the usefulness of the connected approach. The problems mentioned were diverse in nature. However, the significance that was provided by the forum members towards an inclusive approach of design and evaluation of new solutions, the citizen's voice, the patient and the first professional response, were all products of special attention.

I.7. Structure of this book

Books regarding health usually tackle pathologies, therapeutic solutions and consequences in terms of public health. Books focusing on technologies for health emphasize the methods and tools which allow the projects to be carried out and evaluated. This book strives to focus on the usage by underlining the practical difficulties of this "citizen connected healthcare". This choice drives us to propose the following structure, in four parts:

The first part, "Ambitions of Connected Healthcare", proposes a citizen's perspective, in its various aspects: ethics, engagement and participation of citizens and patients, at the individual and social scale. It also addresses the impact of this new healthcare approach on first response professionals.

The second part, "Observations and Measurements", focuses particularly on the transformation of interactions between the patients, professionals and the digital healthcare system. New forms of observation appear – the patient or citizen themselves, and also on the signals they can transmit from a distance – whereas certain traditional investigations are no longer possible. Patient motivation, the compatibility of tools with day-to-day life and the limits of medical knowledge determine the new solutions as well as the potential and technological limits.

The third part, "Methods and Tools for Facilitating Appropriation", develops different approaches for designing and evaluating the solutions, all the while underlining the necessity of mobilizing diverse fields of knowledge to result in desirable and reliable solutions and guaranteeing the protection of the people while remaining economically realistic.

The fourth and final part, "Perspectives", provokes reflections originating from different scientific fields: medicine, social and human sciences and engineering. Alongside these targeted developments, concerning the future of clinical trials and prevention, a panel of experts develops a critical and multidisciplinary view on the feedback of experiments contained in this book.

I.8. Acknowledgements

I would like to extend my thanks to the following for their proofreading and review comments: Véronique Lespinet-Najib of ENSC-Bordeaux INP, Anne Jacquelin of La Fabrique des Territoires Innovants, Paris, and Pierre-Yves Traynard and Helena Burgerolles, both of Pôle de Ressources en Éducation Thérapeutique du Patient Île-de-France, Paris.

My appreciation also to all co-authors and Florian Le Goff and Régis Senegou, both of Société Sephira, Paris, for their participation in the Forum LLSA working group.

Part 1

Ambitions of Connected Healthcare

Introduction to Part 1

The ambitions of connected healthcare, which establish a long-term link between the healthcare system and the individual in their environment, first justify focusing on people. To illustrate this point, we will briefly present two situations: the use of a mobile application to improve the prevention of cancer recurrences, and that of a pillbox which limits the output of toxic medications. These two examples of "connected healthcare" show how it is possible to consider a new way of delivering home care with significant consequences for individuals and organizations.

– Cancer recurrences are sometimes symptomatic. This is especially true in the majority of lung cancer cases (80–90% of cases)[1]. However, in most cases, the patient suffering from symptoms waits many weeks before consulting a doctor. Nowadays, more and more patients are connected: the weekly electronic reporting clinic associated with algorithmic analysis has shown many clinical benefits for its patients, similarly to those evaluating drugs.

– Until now most cancer treatments were administered in hospital, mainly intravenously. In 2020, 50% of anti-cancer treatments will be taken orally (23% today)[2]. However, the systematization of outpatient care leaves patients and medical professionals from impoverished cities facing these treatments and their secondary effects. Conditioning is not adapted to the city. The toxicity of these drugs requires there to be the least amount of handling possible. The connected pillbox could bring some answers to this issue.

1 MoovCare case presented in Chapter 6.
2 Connected pillbox case presented in Chapter 7.

Without minimizing the challenges which these solutions must solve in terms of technical performance, we will concentrate in this part on the human issues. What interests us here, and what motivates economical players engaged in the co-design and evaluation of uses, is the meaning of this "connection" for human beings who interact with the technical system. Assuming there is no technological determinism, we offer to openly analyze the advantages and disadvantages resulting from such a connection, as well as the ambitions and limitations associated with it.

Living every day with a connected device collecting health-related information raises the question of its appropriation by the recipient-individual. This appropriation is problematic, with high stakes in terms of public health and citizen involvement. Therefore, in Chapter 1, we will open the floor to representatives of citizens and patients. This chapter will also be interested in medical professionals who are in contact with patients, particularly first line support, whose professional activity is equally transformed by this new connected healthcare.

Chapter 2 offers a general representation of a sociotechnical system, integrating remote connections and their associated human interactions. This representation has been used as a basis for questioning the work group and as a reference used in the presentation of cases. These cases are inventoried and introduced in an extremely synthetic way. This presentation shows the great diversity in fields of application and solutions, from prevention to care, and including quality of life. It equally underlines that the connected object is nothing without other components of the technical system which it belongs to, and nothing without the human community using it.

Chapter 3 completely develops two cases: that of the monitoring of sleep apnea (SAHOS project) and that of the E4N cohort, individuals whose health is monitored over many generations and whose monitoring will be transformed by these technologies.

In this case, the aim is to give a more detailed report on the interdependence between the components of these solutions and the value they bring to their users, and thus to give a first glimpse of the possible problems which will subsequently be developed.

1

Ethics of Connected Healthcare: the Connected Individual

In the introduction, we touched on the meaning of technical systems aimed at improving healthcare for individuals. Referring to ethics, we are going further: we want to question the "good" effect that these solutions claim to bring to the target individuals. This good will not be obtained despite them. However, the terms "management of change" and "acceptability" tend to take for granted the fact that connected healthcare solutions will prevail for excellent reasons, and that they must and to be accepted by citizens. The approach explored in this chapter challenges this vision, by instead placing the individual's experience at the forefront.

1.1. First approach: the connected pathway

Currently, the expectations of citizens concerning technological solutions are reinforced by the accumulation of digital experiences, undertaken by the users in very different areas: hospitality, mobility, well-being and so on. This accumulation generates permanent comparisons between services which sometimes create expectations of very high standards in certain users' minds. Although they are confronted with a healthcare system, these users will not be able to refrain from comparing it to this multisectoral standard. Other users express their expectations through negative feedback, what they do not desire among the technological offers which they are aware of. The fear of a disclosure of sensitive and personal data has recently been reinforced by

---

Chapter written by Caroline GUILLOT, Jean-Baptiste FAURE and Robert PICARD, with contributions by Myriam LEWKOWICZ.

various cases, similar to that of Cambridge Analytica which led to Facebook being questioned in April 2018. Due to these cases, public perception will consequently be strongly influenced. However, other users will not be aware of such cases, falling outside the "digital divide".

Medical devices and connected objects are intended to be progressively placed at the heart of patients' healthcare service experiences. They fit into a more complete service: hospitable service, hotlines, applications, websites and so on. These services, medical devices and connected objects form the group of platforms used as support hardware for the patient's experience. This experience fits itself and derives all its meaning from the physical, mental and social ecosystems linked with the patient's health, as well as their physical and mental well-being.

The patient always joins an experience and not a medical device or a connected object, which is merely a form of support and must ensure and amplify the experience. Successful services, even if they rely on an object, offer first and foremost an experience.

The patient experience will equally depend on the service's values and its human or relational, even communal, scope; personal causes or the relationship maintained by the individual with their ecosystem: whether it concerns involvement in medical research or joining a patient association and participating in hobbies. These personal causes will contribute to the obtainment of a result.

The notion of a "healthcare pathway" is a new concept underpinned by the development of chronic pathologies and the knowledge that we must develop from a system optimized for treating (and curing if possible) acute crises to a system able to accompany more and more individuals with health issues through time [HCA 12]. Indeed, the patient affected by a chronic illness lives with it permanently. If possible, we must prevent acute health incidents caused by the illness or its treatments and treat them very quickly when they occur. The notions of permanent monitoring, real-time self-monitoring and physiological data (blood glucose for a diabetic individual, for example) are of grant value. It must be possible to evaluate one's state as much as necessary, and to find responses to questions exposed at any time by the "routine" [VAN 15] of everyday life faced with a chronic illness by the afflicted individual. The "healthcare pathway" becomes a sequence of processes marked out through contact with healthcare professionals, but between these steps, an effective communication system is required to answer the patient's questions when faced with unexpected situations, detected by their connected objects. Healthcare and connected objects fit into this evolution insofar as there is a perceived interest in a documented monitoring of these individuals' health status.

It consists in bringing solutions to patients at key points during the experience. Today, the more complex the healthcare pathway (stakeholders, multiple locations), the more the journey is fragmented and becomes a source of anxiety. Between each contact phase with a medical professional or establishment, there is a high risk of "pain points", painful episodes. These consist of difficulties experienced by the individual, psychological or other, and anxiety due to illness. It is these journeys, often experienced during chronic illness or disability, which must be improved first and foremost, focusing the approach on the patient's experience. Connected devices and objects bring a promising element to the extent that they facilitate the monitoring of the patients' conditions and the creation of automatic alerts sent to medical teams and patients, in case of the worsening of the condition, even slightly. The effect of prevention and adaptation will at best be a reduction in hospitalization. However, these monitoring systems will increase the need for counseling and therefore communication between patients and healthcare professionals. The "healthcare journey" then becomes a form of support towards the establishment of new partnerships between patients and caregivers.

These approaches, mobilized in particular by the Living Labs and referring to "Design Thinking" for the design of these devices, go further. This method is based on the issue of incorporating individual experience. The aim is to follow their pathway at the same time as the design phase of tools: the designer looks for points along this pathway which could cause problems from the point of view of the person who experiences the situations. The ambition of the project then goes beyond that of an observation for purely therapeutic and monitoring purposes, destined for professionals in charge, to integrating concern for the individuals, completing the offer of medical value by making an offer to the patient and potentially those surrounding them. This also allows us to remove a metaphorical brake on the appropriation of devices, that of not identifying the benefits of its use. On the contrary, if a benefit is perceived, whatever it is, and whether or not it has a direct impact on health, the device will be used.

To boost this type of project, it is important to start by presenting the problem to be solved, then immersing ourselves in the patient's life and building with them an experience, before defining the technical solution and the "object"/objects to be connected (see Chapter 8).

1.2. The citizen's and the patient's stance

1.2.1. *A connected object for who, why?*

1.2.1.1. *Semantics*

When we consider the vocabulary used while presenting the purposes of an object, we can see where it is and whether or not it follows the traditional care approach. Beyond the fact that these words are consensual in their use, when the vocabulary is comprised of words such as compliance, supervision, education, support, prescription and examination, the underlying professional attitude seems to correspond to an ancient vision of the patient. Submission, obedience and passivity were prevalent before the French Law of 2002 which acknowledged the patients' rights. Words such as: appropriation, self-management, freedom, projects, time, everyday life and family appear within the vocabulary. This changes everything. With this in mind, the patient's interest must be taken into account as early as possible. The patient must be presented very early when it comes to making decisions associated with objects.

1.2.2. *The citizen, the patient and connected objects*

Connected objects are for the citizen and the patient and, despite their technical value, they are one of the many tools among others in the toolbox. The most important is located elsewhere. The value for the individual depends on what we want to solve, on what we will bring. In certain cases, an advanced technology is not the best answer (see Box 1.1).

The Secret Gift project supported by the Encapa association is a board game based on an imaginary medieval fantasy, in which the heroes are "pseudo-diabetic" and cooperatively fight food monsters. The card game is destined primarily at adolescents affected by type 1 diabetes, in order to promote ownership of their pathology and allow them to converse with family, friends and healthcare professionals. Discussing one's pathology with loved ones is indeed not always easy, especially during the complex time which is adolescence.

Secret Gift was a digital project at first, whose objective was the development of a "serious game" used for training young chronic patients relative to their pathology. In the end, Encapa noted the importance of physical presence, contact, in the support of dialogue, and the proposed game became a physical card game. Addressing an issue with a digital solution in mind creates a risk of introducing bias: this resembles the story of the man who handles the hammer and for whom every problem is a nail! One of the reasons for the abandonment of a digital Secret Gift was that trans-generational relationships had to be considered, for which a digital solution could be an obstacle.

Box 1.1. *The Secret Gift project*

A question of similar importance must be clarified concerning connected objects: in whose service do we put them? This point is illustrated in the following.

1.2.3. *The case of observance*

In what way does the introduction of connected objects to citizens and patients contribute to a change of habits among them? At this stage, we offer to illustrate this question, focusing on the notion of observance. A preliminary definition of observance is that of a situation in which the patient understands, accepts and therefore follows their treatment.

Specifically, observance can be seen as the degree to which the patient follows treatment prescribed to them by respecting dosage and set times. This requires an implicit agreement between the patient and the professional, which causes the patient to commit themselves. Observance is a function of this commitment. Knowing to what extent this commitment is negotiated is not always clear, and the asymmetry of information between the patient and the professional must be sufficiently considered for the patient's commitment to make sense to them and to correspond to taking responsibility. Mispelblom Beyer [MIS 16], in his work on therapeutic adhesion, criticizes this concept of observance which he qualifies as paternalistic, and offers instead "therapeutic adhesion", which corresponds to a case in which the patient is an actor and "cares for themselves".

There is a fight around this semantic change. The term observance refers to a point of view in which the patient is inactive, in other words, submits. On the contrary, if they adhere, this assumes that the patient has understood and committed themselves. There has been an active discussion, so that the patient understands what is happening. There is an agreement between the professional and the patient, who accepts the treatment, which comprises a form of acceptance of illness [THI 15].

> Factors of non-observance are diverse and linked to personal psychological evolutions of ill individuals and their capacity to act relative to their health, faults in communications between patients and professionals, the proficiency of healthcare professionals (the best treatments at the right time, etc.) and their practical context, the organization of the healthcare system, access to care [OST 05].

Box 1.2. *Non-observance*

There is always a problem of observance concerning all pathologies, specifically chronic illnesses. According to the World Health Organization[1], 50% of patients, all pathologies considered, do not follow their treatments. These statistics concerning non-observance empirically show that it is an issue of public health.

In this regard, the "need" for a connected pillbox is clearly identified by healthcare authorities; as such a device is likely to become a tool for improving observance. Not taking one's medication is expensive: 9 billion euros in France, for example. But which of the patient's needs does it answer to? How will new uses for this object be perceived when it becomes banal?

Observance is not the lack of disciplinary respect of the doctor's injunctions, nor it is the blind following of instructions written on a product label. It refers to an adhesion negotiated and integrated by the patient from which the product or service can bring value and under certain conditions. While adhesion concerns the use of an object, it is particularly useful for the latter to make sense to the user. Co-design is in this regard a useful approach. It is the whole challenge of the usage of Living Labs.

1.2.4. *From observance to the domestication of tools*

By generalizing the previous point, an important question is whether the connected solutions made available for patients used in the long-term

1 "Adherence to long-term therapies" [WHO 03].

guarantee expected results in terms of health. It should then be a case of preventing device abandonment by the user. This point is illustrated in the glucometer case studied by the Diabète LAB (see Box 1.3). The domestication of tools equally depends on the way the management of the illness integrates itself into the patient's other activities and their family circle. As Aarhus and Ballegard [AAR 10] have shown, technologies must be able to adapt to different patients, as some wish to separate their condition and others integrate it into their lives. Sean Munson's work at the University of Washington (USA) is interested in this question concerning different uses of technologies and reasons for which patients adopt them [SCH 18]. Many projects allowed him to identify a number of hypotheses [MUN 17] which are often created during technological design, and should be questioned, as this book does. Finally, we will consult the work carried out by De Choudhury *et al.* on the predictability of the abandonment of a quantified self or m-health product, or lack thereof [DEC 17]. The approach offered combines an analysis of the device's use and an analysis of participation in social networks.

Beyond the maintenance of interpersonal relationships, the use of social networks can indeed have a positive effect on behaviors which affect health, particularly conversations with peers, the search for reliable information and possibilities of contacting healthcare professionals [WAR 17].

Staying at home is a major challenge for frail individuals' quality of life and for the efficiency of their monitoring. Nobody other than the individual themselves can truly know what living with a chronic illness is like, day after day. Evaluating an individual's capacity and ensuring their daily activities consist of functional tests and general questionnaires. The arrival of new technologies must be considered as an opportunity to reinforce the patient's role in their acquisition of control [HCA 16]. The use of digital devices and tools is becoming part of the "negotiated prescription" and contributes to the reinforcement of integrative practices. Research on this subject confirms that patients' needs are very variable from one individual to another, in terms of support which advocates systems capable of offering a variety of applications to adequately answer to their needs [DHI 16].

1.2.5. *Living well with connections: control or new found freedom?*

A certain discussion promoted by some of the medical profession and by certain industrialists suggests that an ideal part of prevention and care would be a constant monitoring of biological and behavioral parameters, allowing us to know everyone's health status at all times and in all places.

The patients' willingness to provide a form of self-monitoring seems partly conditioned by the possibility of controlling one's care. All of these tools require the ability to understand, use and manage data, involving personal aptitudes and health literacy. There is a new challenge of appropriation arising from educational activities and support for patient engagement.

Additional ethical issues then arise when data is shared by other individuals, and not only accessed by the patient: these issues concern the protection of privacy and the possibility for the individual to disconnect, to quote only two. But another issue is the discomfort resulting from the use of a measuring device, depending on the duration of the observation process. While this concern is taken into account – and we hope this will often be the case – it prohibits certain measurements from being taken or limits the number of measurements, or the time during which they can be continuously carried out, due to the patient acceptability limits. Due to this, compromises must be made: for observation to be acceptable, we must relinquish the ability to provide the most useful parameters. Sometimes, a technological innovation can improve a situation (for example, the measurement of blood glucose). But ideally, the recording of data should be naturally integrated into everyday life.

1.2.6. *Passivity versus activity*

The attitude of each citizen regarding their health, their illnesses when they occur, the professionals and the healthcare organizations, is fundamentally personal and varies depending on context and time. Certain individuals are, because of their history, their age and their temperament, likely to take control, to engage, and to acquire a new understanding and know-how to reduce the asymmetrical knowledge that exists between the professional and the individual or to familiarize themselves with the topic at hand. Other individuals, or the same individuals but in periods of greater vulnerability, will eventually choose, for a limited time, to "let go" and to rely on others. This leads to techniques that are different to technological solutions: some are only useful for a short period of time or for a portion of the population. It is therefore a question of respecting this position rather than imposing it via technological tools. The challenge for these new technologies is to allow the individual, based on validated operational methods, to find their place, depending on the tool used, in the construction of their path towards the prevention of illnesses and the maintenance of good health. This distinguishes the "health pathway" from an industrial process, to which it could be sometimes tempting to bring it closer, all the while guaranteeing that the accessibility of these new methods implies the experimentation of elementary, simple devices at the service of "doing": enriching, creating, producing and repairing. The development of such devices should be conducted with respect to an ethical framework.

1.2.7. *Engagement*

Patient engagement has been viewed for many years as one of the promising ways to improve population health and the problems that confront healthcare systems, via a partnership in terms of treatments, organization of services, governance as well as the healthcare system.

1.2.8. *The active patient*

A recent urgent matter expressed by numerous patient associations is the passing from an "alibi" patient to an active patient. Therefore, for example, World Diabetes Day, celebrated every year on 14th November, was in 2016 made a theme of high priority. There is a shared will in the representation of these associations to bring the active patient forward. However, questions remain as to whether allowing the new representation to become fact, in the concrete medical worldwide organization, should be undertaken. This evolution includes a cultural aspect of which the transformation takes time.

Patients describe themselves as being proactive in three types of practices: continuous learning processes, the evaluation of the quality of their relationship and the degree of reciprocity of the partner engaged with the medical professional [POM 15].

1.2.9. *The implications of knowledge*

The volume of knowledge accumulated by medical professionals, and specifically by doctors, is considerable. Our knowledge concerning how the human body works, as well as illnesses and the ways to treat them, are undergoing strong growth. In addition, the perspectives offered by "Big Data" only amplify this phenomenon. However, from time to time, the knowledge originating from ill individuals tends to be expressed, takes shape, is shared and developed. Empirical knowledge arising from trials and the experiences of lives led with chronic illnesses is specific because it is assembled backwards when compared with the medical knowledge that is based on knowledge established from experimentation. Knowing how to create, in a constructive manner, lives with chronic illnesses so that they can be cured means giving oneself the means to build "from" rather than "react to" [SOL 15].

Considering the patient at the center of thought, "never for us without us", means giving them a position as co-builder in the joint development of a culture of usage related to health. The digitization of civil society, the capacity of connected citizens affected by chronic illnesses to share their experiences, is without a doubt an important lever for change.

1.2.10. *The expert patient*

The concept of the expert patient was first mentioned in English-speaking countries in the 1970s, before being picked up once again a couple of years later during the global fight against AIDS.

Therefore, under the influence of patient associations, lay knowledge is making an appearance in universities. The first initiatives were Canadian, however other countries, and notably France, value this "expert patient", of which knowledge is evaluated and recognized, as well as shared, not only with their peers or their professional interlocutors, but also with professionals recruited from an academic setting.

Patient engagement in therapeutic education programs, and then the actions they perform towards other ill individuals, has aroused the interest of public health authorities. This interest has led these authorities to integrate them in their decision-making bodies and advisory boards, so as to involve the sick in the decisions regarding health that concern their illnesses.

DEFINITION 1.1.– Expert patients can be defined as people with a chronic illness that are constantly learning about their illness and are helping to improve its treatment and/or prevention. They solicit skills that are in the spectrum of autonomy, with some being intercultural and emotional, and these individuals are keen social contributors that subscribe to an ethics based on efficiency and the usefulness that inspires them to contribute to innovation [GRO 14].

Thanks to different initiatives, in France, the notion of an expert patient is starting to emerge. They are progressively becoming the logical sequel to therapeutic education programs that were established several years ago. However, these patients must benefit from a specific course to become expert patient peer helpers and partners of healthcare teams.

Faced with this need for an educational background, some facilities integrate expert patients in university careers based on therapeutic education (for example, within the Université des patients – Paris 6) or in the training of medical students (the experience of Paris 13 University).

1.2.11. *Beyond knowledge: the search for influence*

While the expert patient is recognized for being the custodian of formalized skills that are communicable to third parties and that concern their illness, some expert patients will go further by putting themselves in situations where they can professionally intervene in technological projects by combining their abilities with those useful to the design and elaboration of new solutions. Research laboratories that have staff engineers with higher education, all the while having disabilities that affect their everyday life, have already existed for several years now. However, the movement is extending its reach to people that bring pathological dimensions to research studies, design and development. These are the challenges of knowledge and the contributory value of the militant patient.

1.3. Two illustrations: La Fédération Française des Diabétiques; the ENCAPA association

Two examples are proposed to illustrate the patient's possible position: the first concerns the diabetic, the daily management of diabetes being particularly taxing and requiring an active participation on the patient's part;

the second presents a type of non-specific militant engagement with a specific pathology, simultaneously mobilizing their professional abilities recognized by the market and lay expertise. The technologies that are destined to be given to individuals require their contribution, either in the case of users or in the case of co-designers, so as to make sure that these solutions are the most adapted to their needs. Before delving into these examples, we propose a few problematic elements regarding this patient position.

1.3.1. *The active patient: example of La Fédération Française des Diabétiques*

The *Fédération Française des Diabétiques* was created in 1938 and is a French patient association. Most of the members of the executive board, even the president himself, are patients suffering from diabetes. The headquarters are in Paris, and its national coverage includes approximately 90 associations that are France-based. It has been recognized to be of public interest since 1976 and is approved by the Ministry of Health. It recently became a Federation.

As a player in the health sector, the fédération is fully engaged in the governance of health and the fight against the constant rise of this condition. It carries out different missions, namely informing, preventing, supporting, defending, and financing medical and scientific research:

– the mission of informing is complex. Communicating comprehensible messages to all individuals is a strenuous task as the information is sometimes complex;

– the mission of prevention is very challenging. The national diabetes prevention week takes place every year in June. It is a good opportunity to conduct screenings, thanks to the completion of specific surveys;

– however, currently, prevention is not a task that is sufficiently taken care of by health actors, even though it interests l'Assurance Maladie, the Fédération and the Diabète LAB;

– the mission of peer support is notably being conducted thanks to expert patients that bring great aid to this field;

– there are many cases where individuals wait to be defended, and the legal department of the Fédération answers these calls. These cases can include: assistance with home ownership (access to housing), fights against discrimination in access to employment, to driving licenses, etc.

1.3.1.1. *An issue of health democracy*

The Fédération has launched two initiatives:

1) a training course for expert patient volunteers to accompany their peers during the care period of their illness (in the context of the Elan solidaire project; AFDET training certificate[2]);

2) the Diabète LAB, a tool of the Fédération Française des Diabétiques created in 2015. This was the first French Living Lab exclusively dedicated to diabetes with a community of more than 2500 diabetic participants.

On the one hand, it is a matter of putting an end to "alibi" studies, a sort of "marketing reassurance" for manufacturers who await validation of overall patient satisfaction and, conversely, of developing the ability to analyze patient data and expertise experience, co-evaluation and co-design of new health solutions.

On the other hand, the goal is to reinforce the Fédération's expertise thanks to the experiences undergone by patients to improve care and quality of life. It is a question of passing from testimonies to the realization of sociological studies (see Box 1.3).

This is a large-scale issue because diabetes affects 4 million individuals, of which 700,000 ignore their illness. Knowing as much as possible about them is vital. Each case is different: one patient, one illness and one life path. At the same time, it is also necessary to make all individual voices heard.

1.3.1.2. *The Diabète LAB*

The Diabète LAB is a Living Lab of the Forum LLSA, which follows an ethical and methodological approach. It supports manufacturers: the priority is to move their marketing/sales problems towards utility as well as public

2 Association française pour le développement de l'éduction thérapeutique – French association for the development of therapeutic education.

health issues and to incite them to be involved in an approach concerning the co-development of their products so that they better answer patient needs.

The Diabète LAB gathers facts, for example, regarding medical devices, even though it is still a very complex process to carry out such examinations. The projects are usually temporary (tied to market dynamics); therefore, it is important to develop long-term collaborations. Support for several years makes sense and enables an optimal exploitation of facts originating from studies.

Thanks to capillary blood glucose monitors (to recognize the blood sugar level), diabetic individuals extract a drop of blood by pricking themselves on the end of one of their fingers using a self-injecting insulin "stylus". The facts show that patients do not respect the practices recommended by medical professionals, all the while being ignorant of the fact that they are doing this.

The Diabète LAB is interested in the difficulties encountered in the patient's everyday life. Currently, there are no appropriate practices. The important aim is to understand how individuals get a picture of their, and from this, to build with them these good practices. This way of operating is particularly illness instructive. As it happens, the gap observed between medical recommendations and practices do not originate from product malfunctions, but from other factors such as: the unwillingness to publicly acknowledge their condition, the relationship they have with their doctor or relatives, lack of information, etc.

Continuous glucose monitors, recently available on the market, avoid these manipulations. For example, certain monitors offer a measure of blood glucose by scanning, with the individual's mobile phone, a sensor placed on their arm. They provide, other than a regular update on blood glucose, its tendency to either increase or decrease and predict its future fluctuations. Clinical studies show the positive effects of these continuous glucose monitors on the balance of blood glucose levels and in reducing the feeling of dread associated with hypoglycemia. A study by Diabète LAB also shows that they reduce several social, spatial, organizational, material, corporal and cognitive constraints that are typically experienced by diabetics.

These continuous glucose monitoring devices allow for a greater discretion and promote the multiplication of blood glucose level tests throughout the day. Management in front of others and the practice of blood glucose self-monitoring in new spaces (at work, on public transport) are thus more straightforward.

This study has been of important significance for patients. The work conducted by Diabète LAB and by the Fédération Française des Diabétiques has allowed an increase in the bargaining power regarding negotiations in the reimbursement of continuous glucose monitors obtained in June 2017.

Box 1.3. *Impact of continuous glucose monitors on diabetes*

The Diabète LAB also develops collaborations, with the CNAMTS[3], with cooperatives and with insurers. It also conducts studies on its own initiative.

Thanks to the Diabète LAB, the Fédération shows once again its capacity to adapt to a changing society and positions itself as an indispensable actor in diabetes. The studies realized internally, based on protocols, allow us to tackle the heterogeneity of the population; online questionnaires, sociological interviews and ethnographic observations are all common methods used by Diabète LAB. All the data are protected thanks to a licensed host of data regarding health.

1.3.1.3. *A question of methods*

The Living Labs enlists methods from the human sciences and design is gradually taking a position of higher importance. These aspects are the subject of a specific chapter in this book (Chapter 8). At this stage, we should note that:

– in ergonomics, the importance of the "human factor" is discussed on two levels:

- device acceptance while it is at the individual's and more generally the user's disposal,

3 French compulsory health insurance.

- long-term adoption ("compliance");

– the socio-ethnographic approach allows us to address in-depth the reality experienced by individuals in their environment, particularly that of patients suffering from chronic illnesses. The uses that are established around the appropriation of new solutions transform the diverse aspects of patients' work, the characteristics of their reflexivity and the experience of their illness, by introducing a new temporal plan in the relationship with the self, the body and pathology (see Box 1.3).

1.3.1.4. *Adapting the devices to the patients' real life*

Thanks to the Living Labs' approach, different studies, adapted to a specific context, can be conducted for each solution to understand the problems they can encounter in real-life circumstances. It is in the field that we can more effectively observe and understand the reason for mismatch between usage and doctors' recommendations.

Medical professionals who use Living Labs are also very interested, not only because they bring elements that add to clinical studies, but also because they bring information acquired from real-life experiences, which are otherwise inaccessible.

1.3.2. *The militant patient: ENCAPA*

ENCAPA is a new association that brings together eight individuals, some of them suffering from chronic pathologies, with different careers: hospital engineer, designer, philosopher, pharmaceutical entrepreneurs, project manager in a competitive cluster of biology and so on. It was created in December 2016. Its president, Aurélien Michot, is a hospital engineer and he co-founded the association, in pursuit of a project named Secret Gift (see Box 1.1).

ENCAPA has the goal of promoting and enhancing knowledge based on experience. It favors "empowerment" by relying on the Partner Patient model in healthcare innovation.

DEFINITION 1.2.– Empowerment must be understood as the conquest of a power over oneself, the reinforcement of capacities in some way. This should not lead to a parallel demobilization of public or private institutions in the field of healthcare.

The association develops its action according to two axes:

– axis 1: support, engineering, housing of innovative projects in healthcare (high-tech, low-tech) conducted by individuals suffering from chronic illnesses (project management, searching for partners and funding);

– axis 2: advice, engineering of projects, evaluation, courses that include patients and allow them to act as consultants in business projects, for example, in the pharmaceutical industry or in medical technologies, for start-ups, and also possibly for bigger companies.

What is at stake is co-building adapted solutions by demonstrating the social usefulness of the patient while reducing the total project costs.

The patient can intervene at different times; the list below is not exhaustive:

– during needs analysis, the patient can contribute to the definition of real needs – and not only the ones interpreted by parties such as medical professionals, as is regularly the case. It is important that the patient intervenes at a very early stage, and not only by filling in questionnaires designed by others as that would be inherently biased;

– during the design phase, the patient can participate in concrete choices concerning the first prototype, ensuring that the solution is properly adapted to real-life situations;

– during the prototyping and development phases, they can participate in the recruitment of tester patients, with whom they would develop a strong adapted relationship and a good reciprocal comprehension of the questions at hand. They can even contribute to the organization of tests, analyze and return results, and participate in the adaptation of a solution. If the problems require the upgrade of a solution, they can test the adaptation of the solution to needs and to real-life situations;

– finally, during periods of deployment of the solution, they will contribute to preparing the market for adopting the solution and to its evangelization.

1.4. The connected medical professional and the system

The availability of information issued from "real life" contexts is a challenge for health systems, but it is not a new one. The notion of PROM – Patient Reported Outcome Measures – draws near to our subject and dates back to the late 1980s. The perceived results measured and directly expressed by patients without intermediation, more or less distorted by medical professionals or by their surroundings, are recognized by Living Labs. PROMs improve communication between patients and medical professionals and make it possible to follow, for a long period, chronic pathologies. PROMs also allow the individual or population evaluation of the effectiveness of tools used or actions undertaken. Even though it is possible to involve the patient or patient associations in the interpretation of these measures, this is not the primary goal of the PROM approach [WU 10].

The EMA – Ecological Momentary Assessment – approach includes an individual and behavioral dimension. This approach dates from the early 1990s [STO 94]. It is largely diffused in certain medical specializations and also in sociology, for example. However, it seems that this work has been able to diffuse into the methods of design and evaluation of technological tools. According to [DUN 13], this approach presents the following characteristics, which are interesting for our purpose:

– the ecological aspect refers to environments and "real life" experiences, allowing an ecological validation of the results;

– the momentary aspects allow an immediate, instantaneous result without the bias of memorization or recalls;

– the particular characteristics of the evaluation are that the information gathered by the individual can be repetitive, intensive, longitudinal and allows for analysis of physiological/psychological/behavioral processes over time.

As information is entered, and data are collected at the level of the citizen, other individuals will have to utilize or interpret them. It can be the direct reading of this data; but the data collected is usually obtained after the summary of intermediate processes or of analysis at certain thresholds or other forms of algorithms (see Chapter 6). This question refers to the professional's proficiency (the ability to interpret whatever is addressed to them), to workload (the volume they must deal with), their availability and their reactivity (validation and treatment of alarms or alerts).

The adhesion of a professional to a connection they consider as useful and adding an operational value does not solve, by itself, the question of a solution's sustainability. Experience shows that many objects that were considered had to be abandoned. We will discuss later the technical difficulties that are at the origin of these dropouts. The importance of the experiences lived by a patient has been demonstrated in previous chapters. However, we will now discuss the human and organizational problems generated by these technical limitations.

As far as professionals are concerned, based on the cases presented in the working group, it is possible to identify different causes that lead to abandoned projects: deception, seeing the disillusion of a promise that is not kept; the objective data not being there when needed, they are not reliable and do not allow the decision that has to be taken to improve; there is loss of time, in a context where medical time, or more globally professional time, is seen as precious; finally, the excessive importance of digressive activities, demobilizing and costly, of professionals that have to compensate for the failures of technical systems, even though it is not their job to do so.

It seems that there can be an "overselling" of the connected solution, with a proposition of idyllic value that does not show its limits and counterparts, a poor appreciation of the resources that are mobilized for this to work, an absence of systematic vision that does not anticipate the turmoil that an overlap of data from new fields can bring, from a distance, in a short amount of time, with demands of immediate interpretations (see, for example, [BAR 17]).

As for the organizational plan, uncertainty shifts: the role of those in charge of installing devices, making them work, maintaining them and fixing them becomes increasingly significant. This was not anticipated: neither in terms of the daily role of different actors nor in terms of the induced costs (see section 4.3.2, Organizational aspects).

Nevertheless, the "connected health professional" has a promising future, provided that some conditions are respected: these professionals must be involved in the early design. The solution must be adapted to their use, reliable, validated, not binding, ideally allowing time to be saved by performing auxiliary tasks so that the professionals can focus on tasks of higher value.

2

Introduction to Cases

The thoughts developed in section 2.1 refer to the Conseil Général de l'Economie's (CGE) collective report, entitled: "Technologies et connaissances en santé" (Healthcare Technologies and Knowledge)[1]. They highlight the structuring of the field, but they also allow us to explain the challenges associated with participatory design and the evaluation of the uses of connected solutions. It is then possible to clarify the role of Living Labs in this ecosystem and the problems they must solve. This questioning will structure the presented cases.

On this basis, it becomes possible to state the challenges of connected solutions (mobile applications, connected objects). Finally, an inventory will be presented to summarize the different cases presented in various parts of this book.

2.1. Method

2.1.1. *Connection: systemic challenges*

The prospective approach developed in the CGE's report offers an analysis of technical elements[2] constituting healthcare solutions, according to the triangle below:

Chapter written by Robert PICARD, with contributions by Frédéric DURAND-SALMON and Loïc LE TALLEC.

1 No. 2012/11/CGEiet/SG, https://www.economie.gouv.fr/files/files/directions_services/cge/Rapports/2014_03_26_Technologies_et_connaissances_en_sante.pdf.

2 For a technological approach to this representation, see Chapter 9.

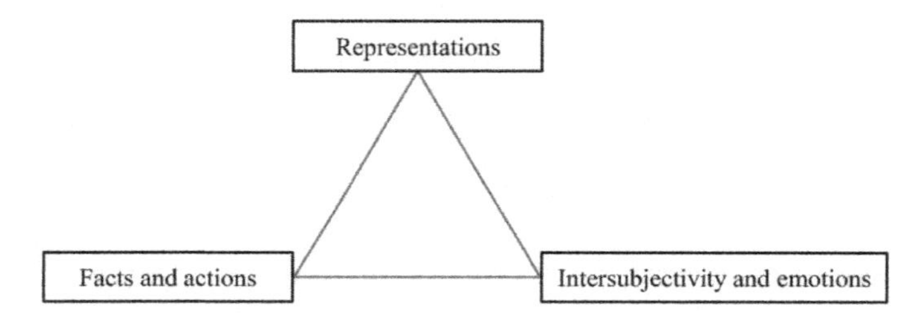

Figure 2.1. *The "prospective triangle" applied In Healthcare*

– the "facts and actions" category consists of objects such as connected objects/sensors (e.g. environmental, physical, biological) and data collection systems (e.g. conjugation of ambient/embedded sensor signals). They have the common feature of interacting with the human body without any voluntary interaction from the individual;

– the "representations" category is interested in elements such as: Big Data (large amounts of data), data mining, inference engines, CDSS (clinical decision support system)[3] and algorithms. They allow the development of new knowledge. At this level, we find the concept of "artificial intelligence" or AI, currently the subject of unprecedented popularity and also a carrier of many risks (see Chapter 6);

– the "intersubjectivity and emotions" category is interested in voluntary and less conscious actions between the technical system and the human, including the emotional dimension. We specifically find coaching applications in virtual universes, serious games and social networks.

The design of elements in various categories, their evaluation, including their interaction is problematic: sensors and effectors at large as well as databases, for example, are often developed by different suppliers.

All of these categories, not only the third, focus on the scope of human interactions, even including the second one which may appear more "technical". The Conseil de l'Ordre des Médecins (French Medical Board),

3 A clinical decision support system has been defined as an "active knowledge systems, which use two or more items of patient data to generate case-specific advice" [WYA 91]. CDSS are computer applications that are designed to help health-care professionals with making clinical decisions about individual patients [SHO 06].

in its white paper on AI[4], reminds us that “medicine will always include an essential part involving human relations, whichever the field, and will never blindly follow ‘decisions’ made by algorithms riddled with shades of compassion and empathy”. This is what sets apart the value of participatory design and the evaluation of uses, and of Living Labs.

2.1.2. *Questioning*

Questions concern the general functioning of a solution. They highlight the necessity of a global and systemic approach to the design and evaluation of this type of solution.

Starting from this model, a first task consists of formulating questions which the Living Labs, specifically, could be confronted with during their missions of design or evaluation of solutions. These questions are the following:

– What is the users’ capacity of interpretation?

– Which data can be acquired in the field? What knowledge is necessary to take advantage of the new system?

– Which measures are possible and useful to implement using sensors and effectors?

– What data will be transmitted in real time to users from sensors? In what forms? How can this data be validated considering the context of its usage?

– What research can be done based on collected data and information? Which algorithms will allow us to validate the experience?

Mobile apps, which are found in some of the cases presented, allow real-time investigations to be carried out. Smartphones are indeed equipped with an internal accelerometer, a GPS, camera, video technology and can be synchronized with other peripheral sensors via Bluetooth (e.g. heart rate, respirators, air pollution, UV).

4 https://www.conseil-national.medecin.fr/sites/default/files/cnomdata_algorithmes_ia.pdf.

In conclusion, Figure 2.2 presents the contributions which these approaches make to the development of medical knowledge and the capacity of medicine to intervene and prevent.

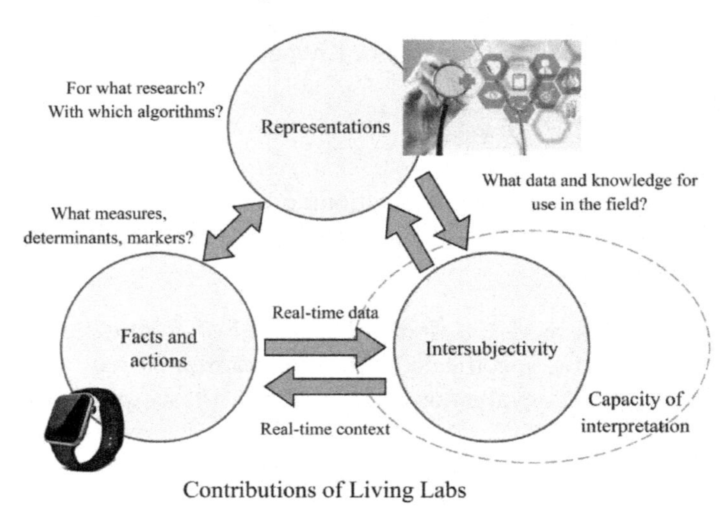

Figure 2.2. *Questions surrounding the design of connected healthcare solutions*

The Living Labs can specifically contribute to a more complete, population and behavioral observation of the public, and to a multidimensional measure of the solutions' impact.

This diagram supports the book's questioning.

2.1.3. *Sharing knowledge*

The collection of data, and its exploitation by algorithms and other statistical processing, allows new hypotheses to emerge – correlations, cause–effect relationships – likely to lead to new knowledge. This knowledge coming from proven correlations, sometimes foreign to human reasoning, can lead to counter-intuitive results. These results will not easily be understood and adopted by field practitioners and patients. This "resistance" is seen particularly when new knowledge requires a transformation of professional practices or lifestyles. The design of information carrying communicating applications of this type will benefit from co-design involving future recipients of new knowledge and information from these solutions. The co-design activity becomes an opportunity to discover and appropriate these novelties.

2.1.4. *Optimizing the organization*

Data originating from sensors can be returned to the individuals from whom they were taken, to professionals in charge of them and to establishments supporting them. They can support and guide collective action, including, where appropriate, the patient themselves and their family circle as actors. The conjugation of data coming directly from sensors with other information input by the patient or a first line professional can accelerate exchanges, optimizing travel and securing decisions. The community of practice in its various components will be mobilized in a useful way when design and global evaluation occurs to anticipate organizational changes and measure their impact.

2.1.5. *Resilience, alerts and security*

For actors to be engaged and trust in the system, they must be sure that the latter functions safely, and allows a degraded function ensuring the maintenance of vital functions in case of possible failures, which must remain statistically exceptional.

The vulnerability of a system lies within its weakest link, meaning there is no reason to use a reliable, robust connected device, ensuring the encryption of information if in other points along the chain or at network communication level, failures may occur which the system would not be protected against.

2.2. Feedback: case presentations

There is a large variety of cases presented in this book. The dimensions of the model take on very different influences. From now on, we offer an attempt at classification, created from a diversified experience of applied projects.

2.2.1. *Typology of applications*

For over 15 years, various e-health solutions have been sold by specialized companies, such as BePatient, based on platforms. Nowadays, thanks to experience gained, different models of client applications allow us to categorize four areas of interest:

– prevention (predictive tests);

– hospitalization/acute care (preparation and follow-up of interventions);

– the management of chronic illnesses/chronic care (secondary prevention, behavioral modification);

– research, using again a collaborative and active approach with the patient.

This segmentation considers the fact that in the definition of products, depending on the area, the user has different designs and usages of a solution. Design plays an important role. In the area of chronic illnesses, for example, social media is important. In the field of hospitalization, it is other functionalities, such as remote monitoring, that are important.

The field of "hospitalization" is not represented in our case studies. The reason for this is that we are focused on connected health in living areas. The elements which follow illustrate three of these four contexts, drawing from BePatient's experience.

2.2.2. *Use case in prevention*

This case concerns the cardiovascular POP (Point of Prevention) solution, which works successfully. It consists of specifying cardiovascular risk in different contexts (general public, events, occupational medicine, etc.). Nowadays, the predictive score involves using questionnaires, biological assessments (which the patient must perform and obtain results of before consultation) and taking measurements. The process is long and requires multiple people, therefore a long time spent for the monitored individual. According to results, the individual receives injunctions which they may or may not follow ("stop smoking!") and without any real solution or continuity in care.

In the case of POP, questionnaires are administered on the platform and integrated connected objects allow results to be collected in 7 minutes. Data come from a sphygmomanometer, weighing scales and a blood sample (biomarkers in capillary blood from a finger). This short time results from the fact that the analysis protocol is very codified in the way it is carried out, as well as the related quality environment (tests, procedures, etc.): professionals are trained to use the connected objects, and suppliers are involved in the quality control of these objects.

The program functions very well in different contexts. Various statistics show why it is interesting to collect data this way, in a real-life context: comparisons have been made between data gathered during a public event in France, data resulting from employee screening in a Chinese company, and others from healthcare professionals at a congress in Mexico. These statistics are very different and suggest that context-specific approaches to public health are coming soon. What is at stake is prevention through behavioral evolutions. It is a very promising approach.

2.2.3. *Use case in chronic care*

This case concerns the co-construction of a program for managing obesity care. It has been carried out with the bariatric surgery service at the Catharina Hospital in Eindhoven.

This construction includes three phases:

– phase 1: verification that the usage of the platform, for the completion of evaluation questionnaires before care, is compliant with what is expected;

– phase 2: evaluation of connected objects and their use, in an independent fashion;

– phase 3: research of the impact resulting from the use of the platform on the healthcare pathway, with or without connected objects.

Concerning the first phase, the patient's feedback is positive: a large majority believes that they have acquired knowledge, they are more confident, they judge the use of the platform as simple and intuitive, and they would recommend its use.

The feedback of the professionals after the first phase shows their interest in pursuing the program: the patients are judged to be "better prepared", the platform helps professionals provide better care and thus improves their reputation. They are ready to recommend the platform.

The results of phase 2 were globally correct (utility, reinsurance, decrease in consultations). Research on the impact of these tools on the healthcare process has been launched.

The positive elements to retain from this experience concern the interest of incorporating the prescriber from the beginning and the relevance of a separate and individual evaluation of the different components, in particular the connected objects, before carrying out the research. Only then can studies on the impact and/or medico-economics be made reliable, but it is a long process.

2.2.4. *Use case in research*

A first illustrative case concerns a program that was co-constructed by an industrialist specialized in the integration of health with a CRO. It specifically concerns a sub-study of the evaluation of a cardiovascular MD, including a study of platform usability to collect patient data using the MD. These data included symptoms and elements of quality of life.

The hypothesis is that data collection in the patient's current life is more relevant than repetitive, spaced out visits, and that a more interactive approach will reduce attrition.

The contents of the application are on the one hand questionnaires on symptoms and on their quality of life and on the other hand elements of "loyalty" (learning – messaging allowing a dialogue with the research – connected objects). A weekly rhythm guarantees optimal use.

Connected objects are implemented, as well as the application allowing the retrieval of measures originating from these objects. Three types of objects are used: a bathroom scale, a device for measuring blood pressure, and finally an activity tracker. The choice was made to choose products of the same brand to ease their integration. When this was not the case, some problems arise: the objects have different designs, and the multiplication of application acts are obstacles to the use of multiple objects. Web platforms are used for training, messaging, etc.

2.3. Working group projects

Summary

The Forum LLSA's working group was provided with many cases in three of the four proposed domains in the previous segmentation. Table 2.1 proposes a synthetic list.

Solution (category)	Population, pathology	Value proposition	Service	Organization	Technique
BYMTOX (monitored)	Post-stroke, Parkinson's	Home rehabilitation	Support, in a playful and continuous way	Reducing the load of PS, autonomy	Virtual games at home and connected
Do Well B. (research)	Autism	Identify real-life stress factors	Measures of stress by digitally quantifying it	Pathological signals measured continuously	Non-invasive sensors and algorithms imple-mented in smart-phones
E4N (research)	Trans-generational group	Heredity, epigenetics and lifestyles	Develop knowledge while reducing the patient's load	Administra-tion of online questionnaires	Connected objects and web platforms
Hadagio (preven-tion, moni-tored)	Elderly individuals, frail	Home support, prevention of frailness	Warnings, care, create links	Professional delegation, autonomy	Secure networks of connected objects + platform
MoovCare (monitored)	Lung cancer	Limit relapses at lower costs	Warnings, limiting unnecessary strain	Reduce periodic tests in hospitals	Mobile app
Connected pill box (monitored)	Polymedi-cated at home	Observance	Accompany and secure the correct intake	Dispensing in the cities of at-risk products	Connected pillbox
Thess	Polymedi-cated at home	Securing dispensation	Monitoring, security	Dispensing in the cities of at-risk products	Connected devices of the admin-istration and tracked

Prisme (prevention monitored)	Post-stroke, Parkinson's	Mobility, reducing falls	Ensure a safe use of the electric RF	Reduced strain (Patient + ps)	Connected electric wheelchair
PTA Santé Landes (interven-tion)	Complex pathologies	Coordination of care of complex cases	Reduce the loss of occasions, reactivity	Organized reactivity, efficiency	Telephone set + objects at home
ADEL Santé (monitored)	Sleep apnea	Following real life, empowerme nt, organized optimization	Inform the actor – patient of their health, alerts, specific to the category of the actor	Collection and aggregation of data from a distance from different sources, documented symposium	Connected mask, accessible database ps = patient = industry

Table 2.1. *The case of "connected healthcare" studied within the Forum LLSA*

This table illustrates the diversity of applications in regard to the concept of "connected healthcare".

Before delving deeper into two of these projects, one clinical (SAHOS) and the other originating from research (E4N), the other projects are briefly introduced below. They will be revisited in the following, through testimonies of those that reported them in the group.

MoovCare

MoovCare™ is a digital application enabling personalized clinical monitoring and the premature detection of relapses and complications associated with lung cancer. The intelligent algorithms that are clinically validated for this medical device marked CE (class I) generate an alert to the medical team in case of anomalies that can translate into relapses or complications – contrary to the standard monitoring that consists of a consultation agreed with their oncologist every 3–6 months, accompanied by a scanner.

Hadagio

Hadagio is a digital platform, developed by the company SESIN. It proposes a package of services that are gradually implemented and that are built around the patient in the context of home support. Hadagio offers prevention, security, coordination, monitoring of care and socialization. It is a global offer that fits into the medico-social domain. Its modularity allows it to adapt to the needs of each patient, and to fit in the beneficiary's landscape.

PTA Santé Landes

The reported experience is one of the introductions to connected objects in one of the French "TSN" programs: an initiative to test digital health solutions in different areas (the Landes). This program proposes digital tools to all professionals for a coordination of the healthcare pathway, as well as a patient tool, allowing both professional and patient information.

BYMTOX

This solution aims to promote autonomy, including at home, intended for the population that are predisposed to having strokes. This objective is more specifically aimed at the stimulation of the upper limb. It is a serious rehabilitation game via mobile support.

Prisme

PRISME allows continuous medical monitoring, in beds and in armchairs. It helps prevent the risk of falling out of bed, out of wheelchairs and during transfer from one place to the other. It incorporates a smart electric wheelchair equipped with environmental sensors that warn the user of potential collision and a medical bed that is equally connected.

Connected pill box

The connected pill box is a connected object considered to be a medical device (MD) in the regulatory sense. The fact that it is connected gives it a new use value in that it is different. The reported project follows feedback from a connected pill box solution.

Thess

The project presented concerns the co-design of a new solution: Thess. It is a solution that involves securing and dispensing oral medication, monitoring administrations and clinical data, in real time, for cancer patients that are treated at home.

ADEL Santé: sleep apnea (OSAHS)

An online business software for the management of the obstructive sleep apnea syndrome (in French: Syndrome d'Apnées hypopnées du Sommeil: SAHOS) is being progressively completed by the collection, on a platform, of data stemming from connected medical devices (continuous positive airway pressure (CPAP) – Pression Positive Continue (PPC) in French). These data are made available to the entire ecosystem according to specific authorizations: doctors, patients, home care service providers, pharmacists, etc. A patient file has been co-constructed by doctors and patients to best meet the uses of the latter. Extension to other respiratory diseases (asthma, COPD) is in progress.

E4N

This project aims to set up and monitor a multi-generation population, by mobilizing digital technologies (the Web, connected objects) for greater efficiency.

Do Well B.

This project aims to continuously measure stress and mood levels in autistic patients, calculated in real time, by the use of physiological signals measured continuously by non-invasive sensors and by algorithms implemented in smartphones.

Diversity and scalability of value propositions and needs

The purposes of the projects are not stable over time. For example, it can happen that the connected object is given a value *a priori* by the manufacturer which is very different from what it will be valued as in the future. The following three examples exist:

– the initial positioning of one of the projects was that of a pillbox to be used in clinical trials to control correct intake. Ultimately, over the course of the project, this pillbox was oriented to be used by the general public, hence the need to understand how such an object would be perceived, so that it becomes an object of choice to improve observance;

– during the first proposition by the company BYM (BYMTOX project), there was a solution of analysis and devaluation of movement. The team at

Saint-Hélier was not interested. However, discussions allowed the identification of another potential need: that of a tool for rehabilitation exercises, associated with sensors, in order to facilitate self-rehabilitation. We have gone from sport to rehabilitation, from the analysis of movement to a serious game, and from solution in the laboratory to a product destined for the public;

– in the case of the Santé Landes glucometer, the transmitter had many assets (it worked with 2G/3G/4G and can connect to other connected objects for a shared transmission of information). However, in terms of use, it proved to be a hindrance because of dysfunctions and a complexification of the process. The experiment was reoriented to integrate, in the connected glucometer, functionality allowing the transmission of information to the platform.

3

Two Stories about Connected Healthcare

To give an idea of the richness and complexity of "connected healthcare" situations, and how tough it is to develop all the cases presented by the group in a single book, we present two of them here. They meet very different challenges, but both fit in the long term. They show that the dynamics of change associated with connected healthcare go far beyond the technological projects' programming horizons.

The first case relates to the long history of connecting devices used for monitoring sleep apnea, and the resulting consequences of the exploitation of the associated data. The second case concerns a population of individuals whose health is followed over the long term (cohort), and to whom connected monitoring is now being offered, thus opening up new perspectives for clinical research.

3.1. Case 1: towards building an integrated solution on the basis of the example of care for the treatment of obstructive sleep apnea hypopnea syndrome (OSAHS)

3.1.1. *Genesis: a medical file meeting only medical needs*

Originally, a small group of pulmonologists developed a piece of medical software for the treatment of OSAHS, on the basis of their needs.

Chapter written by Yves GRILLET and Guy FAGHERAZZI.

For the conversion of a medical file on paper into a digital medical file, the Observatoire Sommeil of the Fédération de Pneumologie (OSFP) allows a very comprehensive collection of content to be maintained, with antecedents, clinical signs, complementary examinations (functional respiratory explorations, polygraphy, polysomnography, biology, etc.), using continuous positive airway pressure treatment (CPAP).

Furthermore, the management of a digital file configurable according to individual use practices requires fewer resources to be mobilized (secretariat) than the management of paper files.

Collectively, the objective of the Fédération Française de Pneumologie (FFP) is to improve, by the use of digital solutions, the effective treatment of OSAHS, which is an important public health issue. Therefore, the FFP decided to establish a national database for OSAHS.

The OSFP is currently used by over 1,000 doctors with differing specialisms. The OSFP includes a total of 108,049 medical files as of 03/05/2018, which constitutes the largest clinical database in the world for this pathology.

3.1.2. *Technical aspects*

The impact of the digital world on healthcare has led the FFP to design a more complete package, the medical aspect being only one of the sub packages. The other sub packages are dedicated to medical devices and the needs of other users: patients, home care service providers (HCSP), pharmacists, etc.

CPAP medical devices used for treating OSAHS automatically transmit their data files to platforms on a daily basis regardless of the CPAP manufacturer and model or the mode of transmission. In May 2018, 200,000 CPAP machines transmitted their data files everyday (observance, leaks, pressure, residual apnea-hypopnea index undergoing treatment).

Each category of users like doctors, patients, home care service providers and pharmacists, uses a specific interface dedicated to their needs which allows them to consult and add to a pseudonymized database.

A scientific study, OPTISAS, whose promoters were doctors (FFP and FSM) and whose principal investigator was Prof. JL Pépin, allowed us to confirm the potential of this type of device.

This study necessitated the collection of medical data from the OSFP, from medical devices (CPAP), from connected objects (e.g. oximeter, sphygmomanometer, accelerometer), from HCSP and from patients (simple self-administered questionnaire completed on a tablet). Decision-making algorithms were used to generate and manage alerts directed at either doctors or HCSP. A follow-up of alerts made it possible to know whether they had been considered and whether an action had been planned to correct them, and the results of this action were monitored.

This study enabled the various actors to see the significant impact of this kind of organizational innovation on each of their professions and the need to clarify the role and responsibility of everyone in this value chain that combines patients, doctors, device manufacturers and HCSP.

3.1.3. *Patients as actors regarding their own health*

Remote monitoring is, by definition, a medical activity allowing a healthcare professional to monitor treatment from afar and to modify prescriptions or care pathways where appropriate.

For chronic pathologies, the involvement of patients, who are experts in living with their illnesses, is essential in order to establish with the healthcare professional, who is an expert in the disease, a negotiated and personalized care contract. It is also necessary for the patient to have the appropriate tools available so as to establish and comply with this contract as well as possible.

It is with this objective in mind that a patient file jointly constructed by doctors and patients so as to best meet the patients' needs was created. A functional prototype of the SomRespir booklet, which is constructed jointly with the FFAAIR, was then tested successfully by a sample of patients before beginning deployment.

SomRespir is the property of the patient who alone judges with whom they wish to share it (prescribing doctor and/or general practitioner or other health professionals). SomRespir can be directly configured by the patient, in particular accordance with the associated comorbidities. This file contains information on the disease and advice approved by the FFP, and tools (scores, visual analog scales) allowing the monitoring of the disease's evolution and the reporting of any possible side effects. It also allows patients to visualize data from their connected CPAP treatment (observance, leaks, pressure, residual apnea-hypopnea index under treatment), whereas up until now this data was only known to the prescriber, the HCSP and the health insurance provider in the event that it needed to be verified. If they wish, patients receive an alert 8 days before their next medical appointment and can fill in their scores and other visual analog scales which can, always with their consent, be integrated directly in the corresponding fields of the doctor's file.

The benefit of data emanating directly from the patient, without any intermediation by a healthcare professional (PROM – patient reported outcome measures), has been shown in many scientific studies[1]. It is then possible to know the patient's perception of their state of health, of the evolution of their symptoms, of side effects and of their fulfilled and unfulfilled needs.

Another benefit of the patient booklet is, not least that it allows the evaluation and/or validation of generic tools and/or specific tools intended for patients afflicted by said pathology, whether these tools are connected objects or applications. It is equally possible to evaluate the efficiency of organizational actions and innovations aiming to improve the quality of healthcare.

Before its deployment at the national level, the SomRespir patient digitalized booklet was the subject of an experiment under the name of Pascaline by the ARS in Auvergne-Rhône-Alpes, France, which was part of the Territoires de Soins Numériques (in English, Digital Health Territories) program.

This successful experiment was based on 300 patients. It benefitted from an evaluation by an independent academic body (IFROSS). The results were positive in terms of satisfaction concerning the usage by patients and by doctors. Economic evaluation is hardest to interpret insofar as it is difficult to distinguish between costs related to the study itself and the costs of the

1 See, for example [WU 10].

patient digitalized booklet. On the contrary, significant economies of scale can be expected from a much broader distribution of this patient booklet.

In this experiment, 100% of the patients who had created a SomRespir booklet allowed it to be shared with the prescribing physician, 75.7% of the patients using the booklet stated that they were satisfied with its use and 62% of the patients using it would recommend the SomRespir booklet.

Figure 3.1 shows a graph extracted from the evaluation report on the SomRespir booklet by IFROSS [PAS 17].

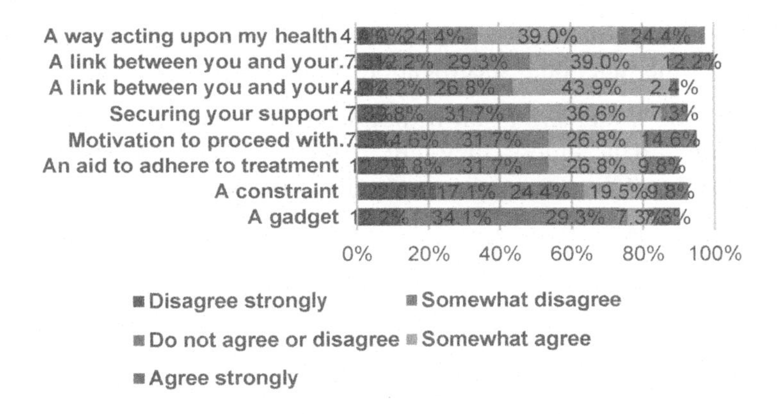

Figure 3.1. *The SomRespir booklet serving the patient. For a color version of this figure, see www.iste.co.uk/picard/value.zip*

3.1.4. *DataMedCare providers of technical solutions*

The company DataMedCare, a startup established in 2012, specializes in the development of digital tools for the medical field and more specifically in the area of chronic pathologies.

In this case, DataMedCare developed the Adel Santé platform. The platform collects data from communicating medical devices (CPAP) to aggregate them. The data thus collected are shared with the various actors of the medical pathway: Adel service providers, Adel doctors (OSFP), Adel pharmacists and patients with a SomRespir digital booklet. The procedures and access authorizations to the platforms were defined by a mutual agreement between these actors.

The functional specifications stated that data must be hosted by an ASIP accredited health data host (the FFP benefitted from a CNIL authorization) and that data collection should be operational regardless of the CPAP machine's manufacturer or model and the mode of transmission used (integrated with the CPAP or an external device). The tools should allow us to ensure the acquisition of the patient's consent to collect their data, to transmit it and to bring it to the doctor's attention. These conditions have been fulfilled.

The partnership agreement between the FFP and DataMedCare can be summarized in a few points:

The FFP grants DataMedCare priority regarding the software developed on the basis of the functional specifications formulated by the FFP.

The FFP will "sanctuarize" all health data. DataMedCare cannot exploit or transfer this data unless they are explicitly requested by the FFP. *A fortiori* DataMedCare cannot commercialize this data.

There is no financial agreement between the FFP and DataMedCare.

3.1.5. *Economic model*

The development costs of the technical solutions (software platforms, etc.), their maintenance and their evolution are covered by DataMedCare.

Public funding will not be a permanent solution so this commercial company must have economic viability guaranteeing its longevity and that of its services.

The originality of SomRespir's economic model in the context of French telemedicine lies in the fact that there is free access for the patient, the doctor and health insurance, and in the impossibility of data being sold.

DataMedCare made the choice to develop a group of services meeting the needs of HCSPs, including the integration of data transmission, and to sell them these services.

In addition to the value in terms of use that have been attributed to it until now by HCSPs (satisfaction of prescribers, anticipated reorganizations induced by the teletransmission in their sector, optimization of technicians' journeys, help with invoicing, facilitation of discussion between doctors and patients), pricing linked to performance introduced by the CEPS[2] should constitute (see Table 3.1) a strong financial incentive.

Observance	Over 4 hours/nights	Between 2 and 4 hours/nights	Less than 2 hours/nights
Remotely monitored	TL1 €17.77	TL2 €16.50	TL3 €7
Not remotely monitored with observance survey	NT1 €15.50	NT2 €14.50	NT3 €7
Not remotely monitored without observance survey	SRO €7		

Table 3.1. *Structure of pricing and reimbursement according to performance*

The differential pricing, according to whether the patient is remotely monitored or not, covers the cost of transmission and the service offered by the ADEL platform. Real-time information regarding the evolution of compliance by patients (especially when it deteriorates and approaches threshold times of 2 or 4 hours) allowing targeted interventions is financially "profitable". In addition, from 2019, the order will require patients to be informed of the data transmitted, which SomRespir already permits.

DataMedCare's HCSP clients are satisfied with the quality of the services provided. They appreciate that DataMedCare is a non-competitive third-party actor and furthermore lets them be independent of MD manufacturers which they used to depend on in order to develop these services.

2 Order of 13 December 2017 modifying the registration procedure and the conditions of support for CPAP medical devices used for treating sleep apnea and benefits associated with paragraph 4 of subsection 2, section 1, Chapter 1, 1st title of the list provided in article L. 165-1 (LPPR) of the social security code: https://www.legifrance.gouv.fr/eli/arrete/2017/12/13/SSAS1735167A/jo/texte/fr.

3.1.6. *What data? For what purpose?*

The data collected from medical files of OSFP are stored in a separate platform. The data collected from patients (SomRespir), CPAP medical devices and pharmacists (Adel pharmacist) are stored in another platform (Adel). Some links exist between these platforms that are hosted by an ASIP-approved health data host.

The data necessary to care for a patient is accessible to the individual and subject to the essential proviso with the doctors they allow to access it. The HCSP supplying their materials (PPC) and the services accompanying these materials have access to their data based on the access permissions that are appropriate to them (see Figure 3.2).

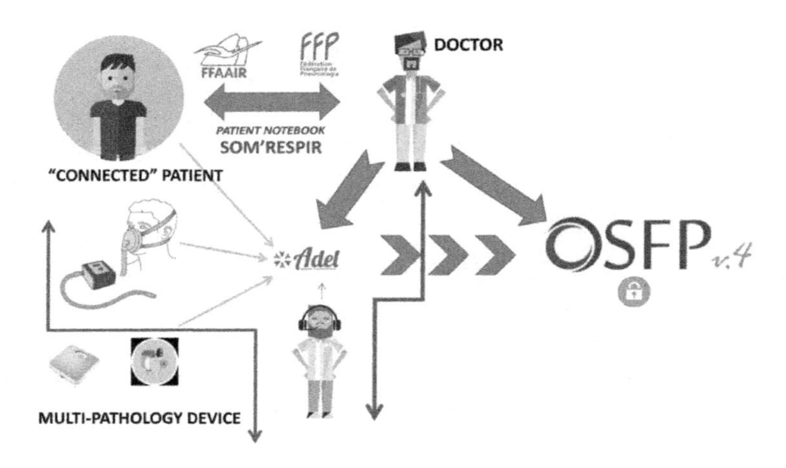

Figure 3.2. *The connected OSAHS patient*

In addition to data chaining for patient care, the volume of data, previously pseudonymized, allows for a "collective" usage. This BDD enables the expansion of knowledge about OSAHS in order to improve its treatment capabilities. Several scientific studies have already been published in recognized international medical journals. Among these studies we can, for example, cite [BAI 16]. This study allowed the identification of six different patient clusters.

In general, the size of the base generally makes it possible to consider the development of algorithms with FFP's medical expertise, by taking advantage of this necessary expertise. French researchers and academics specializing in data, algorithms and artificial intelligence are showing an interest in this database.

Discussions are also ongoing with the Institut National des Données (French National Institute of Health Data) for the purpose of enriching data by accessing medico-administrative data.

The FFP does not have a commercial purpose, and so algorithms derived from these data are intended to be open source. This forestalls the legitimate debates about the market value of data and the distribution of this value among the actors that contribute on a voluntary basis. This voluntary basis is in fact an important guarantee because the various actors can decide at any moment in time to end their voluntary participation (patient and doctors) or their paid participation in home care service providers (HCSP) by stopping the sending or use of their data.

3.1.7. *A prefiguration of an integrated health service focused on the individual?*

For the WHO, the concept of an integrated health service is defined as the following:

DEFINITION 3.1.– Integrated health services focused on the individual means the management and delivery of high standard health services that enable the individual to benefit from monitored services ranging from the promotion of health and the prevention of disease, to the diagnosis, treatment and care required for illnesses, as well as rehabilitation and palliative care, at different levels and in different health care settings within the healthcare system. These services must be affordable, as well as accessible, available and acceptable for whom they are intended.

The exchange and aggregation of data from doctors, patients and medical connected devices by means of ICT in association with the shared medico-administrative data make it possible to prefigure an appropriate tool for achieving the goal of an integrated healthcare service.

In the case of OSAHS, this integration of services offers promising prospects supported by several scientific studies and an evaluation of health records.

The analysis of data enables, among other things, the improvement of knowledge about illnesses, the formulation of decision-making algorithms, and alerts, and at the same time enables us to analyze the efficiency of the care and to evaluate the results of actions undertaken to improve this care.

Connected objects and applications can find their place within this necessary but not sufficient condition that they envisaged, from the outset, with this purpose in mind.

Taking inspiration from the model implemented for OSAHS, the extension to other respiratory diseases (asthma, COPD) is underway with the aim of eventually creating a single portal for all respiratory diseases. It will be accessible to all users, whether they are doctors or patients, who will be able to configure the tool according to their own requirements. For example, regarding asthma, the FFP implements collegial meetings regarding asthma, for the prescription of expensive biotherapies with a registry-type digital medium. Many applications, for example, CNAMTS's Asthm'Activ, and many connected objects exist, so they can be integrated into a more global service. Another example is COPD, of which the most severe forms lead to severe chronic respiratory failure and the need for oxygen therapy and/or non-invasive ventilation (NIV) at home. With the Adel solution, DataMedCare has integrated the list of suppliers declared compliant to participate in ongoing experiments piloted by the DGOS as part of the ETAPES[3] program. In the case of the NIV, the model is easily transposable because the data flows are transmitted by NIV in the same way and from the same manufacturers and service providers as with the CPAP. The only variable is the data.

The currently operational model in the field of pulmonology could be transposed, subject to adaptations of course, to other chronic diseases (diabetes, cardiovascular diseases, renal failure, for example).

3 http://solidarites-sante.gouv.fr/soins-et-maladies/prises-en-charge-specialisees/telemedecine/article/etapes-experimentations-de-telemedecine-pour-l-amelioration-des-parcours-en.

The key factors for success are the application of a Living Lab's fundamental principles, particularly co-construction with the users and the integration of the expertise of all categories of actors to create an efficient value chain. The management of the projects must be based on the inseparable bond that is formed between patients and doctors.

The main difficulties stem from the regulatory inflation which complicates experimentation with innovative solutions, while the concept of soft law is much more suited to the agility needed to remain at the forefront of innovation. In the same way, interoperability, which is a legitimate requirement, is inapplicable in a context where too many public or semi-public actors, whether at regional or national levels, want to impose interoperability but based on a standard corresponding to their own system. The multiplicity of these different systems, not even interoperable between themselves, renders global interoperability, however desirable, utopian.

This concrete realization concerning OSHAS and respiratory diseases offers more ambitious prospects for developing the current healthcare system to be an integrated system that is focused on the individual. For this to happen, it is a matter of urgency that public and private healthcare institutions and public, private and mutual insurers distance themselves from a history of quarrels that are today archaic and misgiving. They have to question their place in the co-construction and co-management of this future system of integrated healthcare that is focused on the individual and incorporates a result-based culture, and whose efficiency must remain a priority.

3.2. Case 2: clinical research and epidemiology 3.0

The case presented here illustrates how the use of the web and connected objects profoundly modify population-based approaches of healthcare for epidemiological research. A cohort allows the monitoring, over time, of a large number of individuals and observation of the evolution of their health status. This number of individuals tends to increase and so does the duration of the monitoring: this should allow us to finely monitor and observe the evolution of the main chronic illnesses as well as the positive and negative effects of health behaviors and treatment.

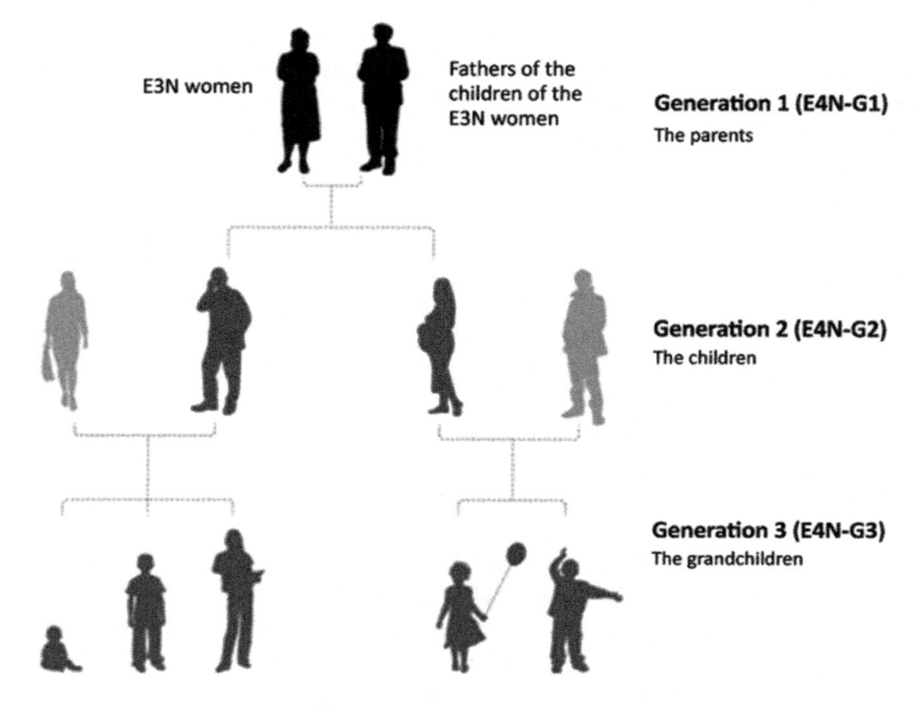

Figure 3.3. *The E4N cohort*

The prospective E4N study was launched in 2011. It is a unique family cohort in its size, design and duration. The E4N project was selected as part of the "Investments of the Future", a call by the Agence Nationale de la Recherche (ANR, French National Research Agency) in 2001. Ultimately, there are more than 150,000 participants spread over three generations: 100,000 women are already enrolled in the E3N cohort study, as well as 20,000 biological fathers of the children of the E3N women (during 2014 and 2015), 50,000 children (in 2018) and 20,000 grandchildren (in 2019).

Among the themes addressed by this cohort, we can mention:

– transgenerational questions: heredity and transmission of health determinants; genetics and epigenetics of chronic diseases;

– expertise on exposure: epigenetics and lifestyle (smoking, physical activity, etc.), microbiota and diet; socio-economic mobility across generations and impacts on health;

– e-epidemiology: integrating new technologies for the collection of information on epidemiology (connected objects, social network content).

3.2.1. *The web platform*

This platform is built as a "Data Hub". It allows the aggregation of data collected from several different sources: questionnaires, SMS, smartphones, biological data and medico-administrative data (SNDS).

Some questionnaires are proposed online, which constitutes a break compared from the voluminous questionnaires diffused every three years by the E3N cohort. These questionnaires are short and concise and adapted to modern life. The questionnaires can be answered using a computer, tablet or a smartphone and some are also transmitted directly via SMS to individuals. A synchronization with connected objects is possible, for example, to gather data on a patient's weight thanks to a connected scale. The objects marketed by Nokia (Withings) have been integrated, but not exclusively; other suppliers will be integrated at a later stage.

3.2.2. *Ongoing scientific projects in e-health*

Concerning connected objects, a pilot study was launched in the E4N cohort, supported by Inserm. A section of the project was supported by E4N, the other being independent of E4N. The latter consists of methodological work based on customer data from Nokia (Withings) objects, in anticipation of a large-scale implementation of these devices in E4N. A study on this methodological approach has recently been published[4]:

– the E4N component of the project involves 700 participants being equipped with the Nokia (Withings) Go device and ensures the automatic continuous synchronization of data with the E4N web platform.

4 For more information, see https://www.jmir.org/2017/10/e363.

3.2.3. *What kind of research question can we address using connected objects?*

Chronic patients and cancer survivors represent a growing population. If we take the example of women after breast cancer, their life expectancy after treatment has greatly improved. Cancer and its treatments can cause physical and psychological symptoms that can persist chronically after the end of treatment. The psychological distress, which can includes depression, anxiety and stress, constitutes a common problem, even years after the end of treatment. It therefore currently seems important to pursue the research in identifying the factors associated with an improvement in the patient's psychological health in order to prevent or at least reduce the psychological distress felt after cancer.

A study on the relationship between lifestyle factors and the psychological well-being of a patient with chronic illness (such as cancer or type 2 diabetes) in the E4N cohort was implemented as part of a PhD project.

The originality of this work stems from the fact that it proposes to evaluate these associations thanks to connected watches, allowing a quasi-continuous measurement of physical activity, sedentary behaviors, sleep and psychological health in a population of 400 women with a history of breast cancer. This project will help refine the available recommendations and implement innovative interventions aimed at the improvement of lifestyle and the psychological well-being of breast cancer survivors.

The Nutriperso project is conducted with the Université de Paris-Saclay, France. It deals with the complex link between diet and the use of technology as a tool for personalized prevention. The challenge is to adapt nutritional recommendations to prevent chronic illnesses based on public health and social and economic considerations. The proof of concept on type 2 diabetes is being funded by the university, in anticipation of the establishment of a larger thematic institute on "Nutrition & Health" at the university.

In particular, this project wishes to adapt dietary questionnaires into a smartphone application. An intervention study will then consist of identifying "high risk" individuals from the E3N and E4N cohorts (biomarkers, lifestyles and socio-economic factors) to identify motivations and factors linked to the appreciation of food (taste, senses) and then work towards a personalized adaptation of nutritional recommendation for the at-risk individuals. A digital support system for the choice of foods will be developed on this basis, destined for the participants, with instantaneous and individualized recommendations being provided. We will evaluate individual changes (biomarkers, risk perception, food choices) and health benefits using the digital tool after 6 months of use.

More generally, in the short term, a "connected" E4N sub-cohort will be created: on the one hand, spontaneously by the participants themselves if they already own connected objects, and on the other hand, based on funded projects, thanks to which the research team will purchase connected objects and offer them to the participants. The objective is to acquire additional information unavailable through traditional questionnaires. Crossing declarative data and objective measurements will allow a more reliable characterization of the participants' behavioral factors. There is an increase in the number of connected objects among the population, which leads to a change in the way that epidemiology is undergone. It is no longer just about making an observation, but about developing the interventional research of tomorrow, by changing lifestyles thanks to real-time feedback from individuals.

PART 2

Observations and Measurements

Introduction to Part 2

The use of the term "connected" implies the existence of technological means of connection. There are promises associated with this word, "connected": capturing or inputting information, allowing adapted individual or collective actions or reactions as well as the development of knowledge. This having been said, the aim here is to understand how and to what extent these promises are realized in practice.

To understand the aspirations and potential deceptions, it is necessary to tackle issues such as the relationship between the imaginations of users, citizens, patients and professionals and the representations of the designers, on the one hand, as well as the theoretical and practical limits of implemented devices on the other hand. This gap influences the perceived value of these solutions: it must be based on facts and must therefore be measured. The result of the measurement is often meant to be shared. However, this sharing does not mean that the resulting knowledge is equally distributed. The introduction of connected devices leads to a modification of the relationship between actors, as well as a potential evolution of their roles and responsibilities: these aspects are covered in Chapter 4.

In Chapter 5, we will focus on the devices that are connected within the real world and allow the collection of data, with or without human interaction and according to different time scales.

Chapter 6 goes back to the processing of massive data: artificial intelligence and algorithms, the potential and limits of which will be briefly presented.

4

Measurement and Knowledge in Health

Connected healthcare implicitly brings the question of measurement: monitoring a patient's state of health, generating an alert or alarm if the situation deteriorates, asking the patient to enter information about their health, all involve collecting health-related data, and therefore measuring it. After having discussed this notion of measurement, in this chapter, we will focus on the different measurement methods: automated versus acquired by a human. We are reminded that these measurement results become sources of knowledge if they are collected, compared and processed. Finally, using the Santé Landes case, we will illustrate how this knowledge can be shared and bring about new collective dynamics.

4.1. Measurement and knowledge in well-being and health: fundamentals

For millennia, providing comfort and well-being, curing a sick individual and alleviating the suffering of the wounded have been very ancient acts, which are also found among the animal kingdom. Measurement and medicine come from the same roots: magic, medicine, divination, reflection, observation, know-how, measurement and equilibrium, such are the terms which characterize ancient medical practices. The etymology of the word "measure" has its origins in Sanskrit. The *maã* root, "magic and illusion" is transformed into *me**/ in Indo-European. This gives rise to not only the terms medical, medicine and meditate but also *mestizos* and *measure*.

Chapter written by Marie-Noëlle BILLEBOT, Marie-Ange COTTERET, Patrick VISIER, Norbert NOURY, Henri NOAT and Robert PICARD, with contributions by Nathalie BLOT and Bastien FRAUDET

Well-being, health and connected healthcare are primarily governed by measurement, metrology, quality, the available standards and instruments, as well as the search for a middle ground and balance which defines a state of well-being and health.

After having discussed this notion of measurement, in this chapter, we will focus on the different methods of measurement: automated versus acquired by a human. These measurement results which are then collected, processed, analyzed, compared to references and associated with their uncertainty are carriers of knowledge.

4.1.1. *The right measurement*

The concept of measurement which seems so obvious in everyday life comes from complexity and questioning of essentials and fundamentals. In reality, the questions asked, before all measurement activities, regardless of nature, are: what do we want to measure? What is the magnitude which we wish to measure (*measurand*[1])? To do what? With what objectives and constraints? With which tools? With which modes of operation? With what reference, units? *With what uncertainty*?

DEFINITION 4.1.– **Measuring.** Metrology, the science of measurements and their applications, encompasses all theoretical and practical aspects of measurements, regardless of the measurement's uncertainty or field of application[2].

To measure is to compare: it involves comparing an unknown physical quantity with another known quantity of the same type as a reference, using an instrument. It is the expression of the result of this comparison using a numerical value, associated with a unit which recalls the nature of the reference, and is accompanied by an uncertainty which depends on the experiment carried out and the knowledge of the reference and its conditions of use.

Many sources of error affect the gross result of a measurement: the measured quantity itself is sometimes badly defined, varies in time or space,

1 Definition: quantity intended to be measured; Source: International Vocabulary of Metrology – Basic and General Concepts and Associated Terms (VIM 3rd edition).
2 Extract from "International Vocabulary of Metrology – Basic and General Concepts and Associated Terms (VIM)".

or is affected by the act of measurement; sensors and instruments used have flaws where appropriate; the mode of operation introduces errors; many "influential quantities" characterize their ambient conditions influencing the result and so on. Corrections must be introduced to compensate for these errors.

This result is not a definite value: it comes from results presenting a certain "spread", and additionally, there is a certain lack of knowledge concerning the value of each individual correction, and therefore the total correction.

Once all these causes of error are considered, the parameter associated with the result characterizing the spread of numerical values is called the uncertainty of measurement, which can reasonably be attributed to the measurand. Moreover, the "suppliers" of measurement results and their "clients" must express the results using references unequivocally known by all actors and expressing the associated uncertainty using an agreed scientific method. This group constitutes metrology, universal language of science, techniques and society[3].

4.1.2. *The correct measurement*

Socially, a metrological pact between two or more people is needed. In other words: to generate trust by coming to an agreement.

Trust relies first and foremost on "the precious human factor[4]". One's own connectivity in a healthcare relationship is a link to others. As a patient, measures taken by themselves are/will be interpreted most often in the form of graphs by carers upon arrival. If empowerment involves the right to take one's own measures, then understanding the operating mode used to interpret them is another which justifies common metrological culture, as well as a degree of uncertainty linked to measurement results.

3 Marc Himbert: http://www.metrodiff.org/cmsms/index.php/metrologie-contemporaine/cssmenu_horizontal.html.

4 Article by Pierre Vandenheede: https://fr.neurocognitivism.be/particulier/blog/article.php?doc_id=591.

Placing the person at the center of the task of measuring themselves, or even forcing them to be the direct data provider, also involves setting in motion a common reflection on our methods of empathizing and communicating, as well as personal well-being and health.

In connected healthcare, so-called empowerment[5] methods are widely used, and collective and participative intelligence is discovered on a daily basis.

The Canadian "Barometer" or the "Beflow" measuring machine illustrates this point (see Box 4.1).

The *Institut universitaire de première ligne en santé et services sociaux* (http://www.csss-iugs.ca/iupl-equipe) developed the "Barometer" tool. It consists of a clinical, digital and collaborative intervention tool which highlights the strengths and progress of an individual in their community. Starting from dimensions and indicators recognized by the scientific and experimental community, which can be added to by the individual themselves with or without the assistance of a third party, it allows for appreciation of the evolution of quality of life while keeping priorities in mind.

It is based on the following principles:

– personalization and co-production of care and services;

– reinforcing capacity based on the strength of individuals;

– strengthening the individual's choice and control; and

– the promotion of individuals' knowledge.

It aims to be a vector of empowerment in a context of interprofessional practices and is therefore coherent with personalized care projects for the sick or frail individuals.

Box 4.1. *Barometer: a self-measurement tool*

"The body tells a story and the body does not lie". It is according to this principle that Beflow's measurement and interpretation system is based upon. Using six electrodes placed under the feet, under the hands and on the head, Beflow collects various pieces of information on the individual's state. There are 30 primary derivations enriched by an algorithm allowing us to obtain information from 180 areas of the body. The basic principle

5 Other similar terms: potentiation, empowerment, ability to act and power to act. This term is discussed in section 1.3.2.

is to measure the resistivity of these different zones in correlation with the body's pH (see Bioelectronique de Vincent[6]). Each zone can be in a state of equilibrium or in a more or less serious state of acidosis or alkalosis.

This basic data is then reprocessed by a computer server and allows the following information to be obtained:

– the physical state;

– the mental state;

– the emotional state.

This non-declarative measurement brings a more "objective" vision, both to the professional and to the measured subject. For example, we note that more than 50% of individuals who claim they are "tired" are in fact mentally tired and not physically so, which they were rarely aware of.

In 15 years of experience and thousands of measurements made, including a study with the dentistry department of the La Salpétrière hospital and the Faculté de Lyon 2 on measuring dentists' stress, we have found that the communication of objective data alone to the measured individual generates an awareness of certain phenomena and motivation to improve one's state. The periodic performance of measurement on one individual allows them to follow their evolution and reinforces their motivation.

Beflow is therefore a tool of empowerment which fits into the framework of improving individuals' autonomy at different levels[7].

Box 4.2. *Beflow: a system to include "objective" measurement*

Let us add that the field of connected health calls for artificial intelligence[8]. On this topic, let us put together Pierre Vandenheede's question:

> "Why does the human factor remain a good investment despite the promises of artificial intelligence?"

6 https://www.votre-sante-naturelle.fr/.

7 http://www.bybeflow.com/index.php?option=com_content&view=article&id=74&Itemid=480.

8 The theme of artificial intelligence is developed in Chapter 6 of this book.

His statement is such that we could all agree. He adds:

> "Beyond a certain fatalism, to consider man for what he specifically has, can open us up to new dimensions which could, hopefully, let us face the many challenges presented to us and that the news continuously repeats: global warming, extinction of species, commercialization of society, etc. Therefore, and despite the uncertainty which it entails, it seems to us that it is more important than ever to invest in humans to highlight the ways in which they are valuable!"[9]

Linked to everyday common sense, measurement is a tool needed to create awareness. In healthcare, those from ancient civilization thought a good physical balance went hand in hand with mental, affective, social, creative and cosmic balance. Greek doctors and philosophers laid the foundations of the relationship between well-being, health and the environment.

Current measurement devices allow explorations of the infinitely complex world of all living things and the universe that surrounds and protects us.

Connected healthcare is a matter of measurement and metrology which must be shared. On the patient side: being able to enter data about their health which implies measuring their own physiological parameters and therefore giving themselves the means to take (or not) control of their own healthcare's evolution. From the medical body's point of view: from measurement results, giving a correct diagnosis, to following the patient's state of well-being and health, being reactive if a situation deteriorates, following the effects of treatment, etc. Patients and care teams apply methods and use measuring instruments which are more or less sophisticated, with more or less skill and precision.

Recent research in cognitive science and neurobiology shows how everyone unconsciously uses, at every moment, thousands of sensors and measuring instruments allowing us to continuously measure our internal state and external environment. All of this data is in extreme abundance.

---

9 Article by Pierre Vandenheede, *Idem*.

Processing of this data by our brain and body is still poorly understood and yet within reach of all living beings.

From paramecium to the elephant, we can say that the animal and human kingdom cannot live without constant adaptation. We instinctively know whether we are hot, cold or hungry, if we are standing, sitting or lying down, still or moving.

Each conscious gesture, of which an individual can define, represents an opening in the field of consciousness. Newly gained knowledge, whether it is intellectual or manual creates a new set of neuronal connections, bringing new knowledge to the body, and establishes a new balance and new possibilities. Like a digital fingerprint, a personal measurement system is unique.

Personal metrology which we approach with envy and desire allows the escape from one's internal confinement and thus evolves in a larger space, a new perimeter of freedom. Personal metrology organizes, structures and expands our field of consciousness. It allows, through measurement, the marking of a new space. This personal metrology is completed by the results of objective measurements which serve as a basis for an enriched discussion during interaction with a professional, a team. This is a kind of discussion where the connected patient is aware of the importance of their well-being thanks to device acting as a fully integrated member of a care team[10].

4.2. Modalities of measurement

A measurement concerning a person can be done in two main ways: by capturing a biological or behavioral signal without human intervention and converting it into interpretable data, more and more often in a digital and potentially even transmissible form, or by the individual who provides this data. This can result from the reading of a technical device and also from subjective elements and feelings, as well as the individual's emotions (pain scale, for example). In the following, we will examine the characteristics of these two modalities.

10 Marie-Ange Cotteret, extract from *Mesurez-vous!*, 2008.

4.2.1. *Sensors: typology and objectives*

The purpose of measurement is to allow comparisons and to project situations in similar spaces, either with other patients, or with the same subject but at different times. Indeed, the doctor's senses do not provide numerical values! It is therefore necessary to quantify the measurement. Associated with this are requirements of quality, cost and reliability of measurements (human factor). This particularly depends on the decisions that will be made using this data. Originally, we mobilized biological and medical engineering (BME): what are the magnitudes to be measured? (temperature, pressure, acceleration, etc.). A transducer then converts these magnitudes into quantifiable quantities (voltage, force, torque) that can then be displayed. A sensor regroups these last three functions.

DEFINITION 4.2.– **Biosignals.** The variation of a physical quantity within the body is called a "signal" or "biosignal". There are six main categories of biosignals (Figure 4.1):

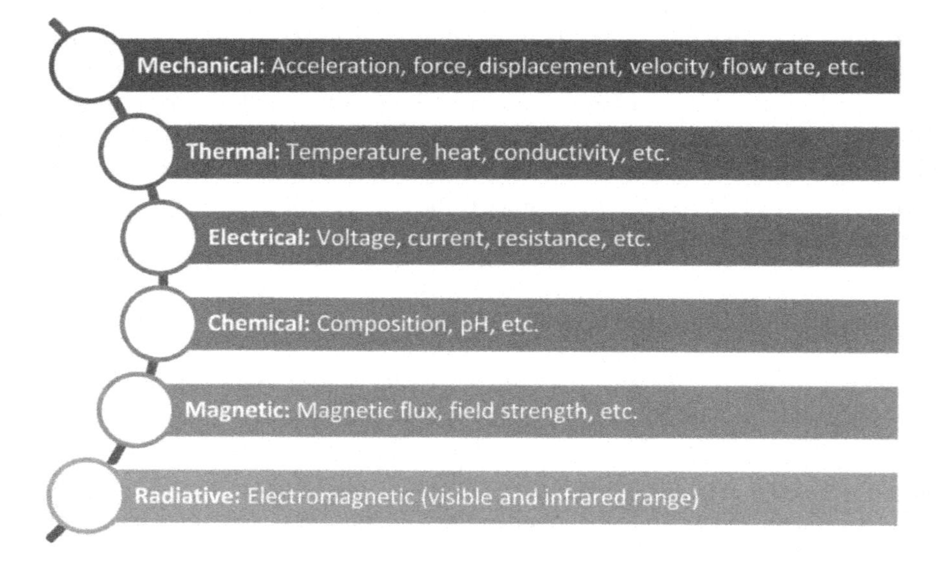

Figure 4.1. *Biosignals (source: INL Lyon[11])*

11 http://inl.cnrs.fr.

BME aims to develop new therapeutic and diagnostic tools for healthcare. This consists of implementing a technological and industrial approach along with many scientific sectors: mechanics, physics, chemistry, electronics and informatics. The industrial sectors which feed BME are varied: clinical engineering, nanotechnology, medical imaging, biomechanics and biomaterials, medical instrumentation, and medical devices. These last two sectors are those concerning the electronics of portable sensors.

There are four types of technological opportunities:

1) MEMS (microsensors) and NEMS (nanosensors), the key features of which are size and cost. These technologies are less than 20 years old and have been popular for the last 10 years. We are within the technological field of microelectronics, with the ionic etching of supports. Microaccelerometers embedded in mobile phones are part of such technology. NEMS will allow for more precise targeting of areas of action.

2) Microcontrollers (microprocessor + peripheral analog devices + time management) – μDSP (microcontroller specialized in signal processing) – built-in high capacity memory.

3) Information and communication technologies, particularly in the global network: Internet, popular use of wireless communications and connected objects: the Internet of Things (IoT).

4) Embedded signal processing. These technologies appeared around 20 years ago. They include embedded algorithms.

The network of sensors on the body is an opportunity: BAN (Body Area Network), including the wireless sensor network, WBAN. Indeed, an isolated sensor is not sufficient: more information is needed, for reasons of either redundancy or completeness. Each sensor provides a piece of information, which must then be merged. We do not create the network because the know-how is available, but because it is necessary.

The personal access point to the Internet is usually the smartphone, which has high connectivity, high storage capacity and integrated sensors, and does not lead to additional cost as most of the time the user already has one on them.

The use of the smartphone as a WBAN portal is needed specifically for the following reasons:

– It allows for a fully interoperable connectivity with WBAN, which ensures long-distance network connection (and remote servers) even in the case of mobility. Wi-Fi and older Bluetooth generations (BT 2.1) are not suitable. Low-power Bluetooth (BT 4) offers advantages for long-term scenarios.

– Furthermore, the smartphone allows a biological feedback for the user, with easy development of various educational and monitoring applications. It provides the framework for the rapid design, deployment and evaluation of applications.

The electronics of portable biomedical sensors fit into general electronic architecture: integrated sensors communicating (using UART – Universal Asynchronous Receiver Transmitter – protocol, Serial Peripheral Interface (SPI), I²C, etc.) with the analog converter, transducer, conditioner, power management and communication module (protocol stack and radio frequency transceiver). The converter comes with event management and digital processing.

A biomedical sensor does not usually have an integrated display. However, it contains the necessary electronics for conditioning and transmitting measurements.

An example is that of Rec@med, a cardiac activity sensor. The monitoring of cardiac activity's temporal variability allows information on the activation of the sympathetic and parasympathetic systems[12] to be gained. Requirements for such a device include reducing bandwidth and providing the user with local feedback.

A WBAN[13] is integrated in a complete system allowing the continuous monitoring of variables sent by sensors. We distinguish the monitoring

12 The functioning of these systems is described in a more precise fashion in the scope of the "Do Well B" case in Chapter 6.

13 Wireless Body Area Network (WBAN), which consists of a network of sensors built to function autonomously and to connect to different medical sensors and applications, situated inside or outside of the human body.

provided by external medical services, accessible via Internet, and local connections established in living spaces, which manually integrate function for the concerned individual.

Ultimately, there exists a continuum of devices and technological systems at different scales to achieve such monitoring: on the one hand, there are non-invasive sensors (proper sensors, bracelets, smart clothing); and on the other hand, there are "exo sensors", actigraphy, tracking and identification devices, which allow the monitoring of circadian rhythms via a "smart home", as well as measures of mobility.

4.2.2. *Interactive applications*

Technical devices made available to the individual or set up in their own environment may subsequently require intervention. These data are to be saved or validated, depending on whether or not they come from a connected object. It can also concern subjective elements, feelings and emotions. In some cases, the mobilization of the individual to carry out these actions can be considered an important or even essential part of therapy. Sometimes the need to make the individual input data is linked to other considerations: for example, available connected sensors do not perform well or are not reliable enough, because they are more expensive than those which are not connected, or not available in proximity of the patient's living space. The multiplication of the connection of different devices that do not share the same standards can prove to be costly and complex.

If connection is not a guarantee of a therapy's success, the opposite is true, and the patient's participation in data collections is not necessarily a guarantee of success (see section 11.1, Chapter 11).

4.2.3. *The Hadagio case*

The case presented in section 4.2.3.1 particularly illustrates different aspects of the measurement problem. It will be referenced several times later in this section.

4.2.3.1. *Hadagio: a package of services for the management of frailty and home-based care*

The elderly yearn to stay at home as long as possible. This is the wish of at least 80% of them. In this context, it consists of satisfying the needs of medical coordination and access to patient information. This concerns not only medical professionals but also social partners who are concerned by frailty, such as home-based carers. Above all, it is indeed about preventing risks of frailty, for example, by collecting information from connected objects. In this regard, it is necessary to have the most precise information. It is also necessary to act on the socialization of territory, which has effects on the populations' health, and to be concerned about the well-being of these people.

The Hadagio solution offers a package of services. The offer concerns prevention, security, coordination, monitoring of care and socialization. It is a global offer which fits into the socio-medical world.

Another challenge is to find an economic model: the chosen approach consists of mobilizing the Web 3.0 model, to create new networks connecting frail individuals to their friends and family (social portal). This must make it possible to finance all of the services.

A modular and adaptive offer

The services offered are gradually being placed and built around the patients as part of home-based care. The modularity of the offer aims to adapt to each clients' needs, to fit into the beneficiary's landscape.

The patient acting on their own health

The solution integrates the concept of the "life project" according to which the patient is called to reflect, in a logic where they become their own health actor, and which enables them to determine their own fragility. Each patient fills in a questionnaire to define their own goals. These goals are then periodically pursued. The solution can be configured accordingly. This can be done monthly, every six months, every year and so on; this varies depending on each case. The questionnaire is therefore provided, in a declarative manner, and the follow-up is adjusted according to the answers provided.

The solution presents available health data whether it is declarative or collected by the means of a connected object in the form of curves.

The choice of connected objects

The solution can integrate diverse sensors: pill boxes, blood pressure monitors, crash detectors, gas detectors, etc. The connected objects are, in terms of needs, selected and adapted according to each patient's specific situation. These connected objects allow the automated collection of data by means of sensors. This is the case, for example, the weight of food. In this example, the solution proposes a placemat that weighs ingested food. The amount of water drunk is also measured. The distance that the individual travels is measured by a pedometer.

An adviser helps the individual to choose the connected devices that are relevant to their utility, which can lead to an integration of objects that are not developed by the supplier market.

Alerts and collaboration

The solution also allows the monitoring of patients' home security and well-being, via the means of a robotic solution. At the same time, it is part of the Health Act: when alerts are generated by connected objects, it is the patient who determines who will have access to what data. It is then the patient who determines who receives information, and to what extent an alert of a given level is received and by whom. It appears that, in practice, it is not always easy for the patient to define and diffuse different levels of alerts. It seems judicious that medical professionals collaborate with the patient to determine who receives what.

Data processing

The measurements that are undertaken are done so either by a medical professional for health data or by the patient for behavioral data. If there is a connected object that automatically takes a measurement, the data generated from it enriches the medical file. The value of the data depends on the way in which they are collected: if it is done by a medical professional, it will be validated medical data; if it is done by a frail individual or a carer, it will be indicative data.

There also exists a behavioral file that provides information on lifestyle, habits, diet, etc. Menus can be based on a person's health factors, on behavioral data collected or transmitted by connected objects, etc. This file more specifically concerns the end beneficiaries who are very vulnerable or have serious loss of autonomy. However, not all end beneficiaries are in these circumstances, and some can simply be health-conscious people.

This data does not directly correlate with information on health, but it brings elements that may have an influence on health statuses. The behavioral file is separate from the health file, but it feeds into it and allows the documentation of health statuses crossed with health data.

The two types of data (medical or behavioral) feed into a health booklet proposed in the solution, including real-time data from sensors.

Communication and coordination

The obtained information concerning the patients allows monitoring, observations, etc. Elements are sent to a module that tracks tasks and facilitates the coordination of medical professionals and social workers or other patient-involved organizations: for aid functions, actors from CCAS – Centre Communal d'Action Sociale (French Center for Social Action; various locations), etc. These actors can then organize their interventions around what happens to the patient. The information concerned is personal intervention, care and follow-up of tasks.

The social portal

The contents of this portal are aimed at personalized services: education, e-learning, therapeutic education, as well as links with the outside world: blogs, forums and social networks. Through this portal, the individual can establish links with their family, their entourage and strangers. The solution is indeed open to the entirety of the population and to individuals who can find each other on the basis of passions or shared experiences. This portal has encountered some success: initial feedback shows that it is used by patients for around an hour and a half every day.

Remote monitoring

If a fall is detected (proven or "soft" falls are included in the events detected, among others), we must understand the context, what happened, in order to decide and guide intervention. For this, we have a webcam, which turns itself on following the detection of the fall and sends an image of the context where the body is pixelated. The distant operator can orient the camera, address the individual and ask them questions to discover more about their condition. This allows detailed interventions. The images that are received are stored in an encrypted form for use in subsequent examinations.

Alert management

For each sensor, a process is created with configurations: Text, Level and Recipient. The alert is received on the recipient's cell phone according to a predefined setting specific to each case (what should be sent to whom and in which circumstances according to which modalities, e.g. an SMS for hypertension). The practitioner can open a window on their cell phone and access live additional information. They can realize a remote consultation, access medical files or the patient's history and create a prescription with Vidal medicine dictionary online. Remote expertise is possible, i.e. multiple simultaneous videos. The interpretation of the value of each sensor goes through a rules engine. A simple rule means that only one condition triggers the alert. A multiple rule means that multiple conditions can trigger the alert – the number of conditions is not limited. A compound rule is a rule where context, history, calculated values, etc. are all considered.

The means of monitoring

A joint project is undertaken with object manufacturers for the development of the connected device. The transmission of data is therefore automatic. This transmission is provided by radio and encrypted between the object and a box in the individual's home, in this case the tablet, by secure messaging to a data server that also embeds a rules engine.

It is specified that it is not necessary to have a high-speed Internet connection, because the daily flow is very slow. The automated recovery of data is available for half a dozen pieces of equipment: in this case, the individual has nothing to do. We note the presence of home automation equipment. Hosting of health data is provided in accordance with regulations.

The Hadagio solution was designed and developed according to a co-design involving future users, patients and medical professional alike. This work was conducted, among other things, to study the challenges and risks involved in integrating connected objects into the solution. These elements, as well as the technological requirements of the solution, are presented and discussed in Part 3, Chapter 10. The solution continues and enriches itself thanks to user feedback.

4.3. Collective intelligence

The gathering of data originating from a human individual, from a group or from a population is not neutral: it is the source of novel information and knowledge and it potentially modifies the behavior of these individuals, as well as their interactions with their entourage and the medical professionals who take care of them. The technical system modifies the human system, and the expected effects will be there only under certain conditions that are not only technical. This section successively addresses some aspects of this problem: the sharing of knowledge, organizational aspects and the reactivity of the sociotechnical system.

4.3.1. *The sharing of knowledge*

The gathering and exchange of information and data takes place in a context that will give them meaning. This context can itself be embedded in the technical solution, and the restitution of measurement results can be dependent on the application and the person. We will illustrate this point with two examples.

The Hadagio case presented in the previous section allows the patient to "measure their frailty" by completing online questionnaires created to allow them to define their own objectives which are then monitored periodically. The follow-up is adjusted according to the answers provided.

The display of data "makes it possible to exploit concomitance of events"[14]. Furthermore, connected objects are selected according to the answers provided in the questionnaire. An adviser will then help the patient to choose connected objects that are relevant to their case.

The BYMTOX case developed in section 8.3 presents a sensor-based solution from the field of sport and movement analysis that aims to promote

14 Blot N., Noat H. (SESIN), Presentation, Forum LLSA working group, Paris, October 27th, 2016.

independent exercise, including at home; more precisely, the stimulation of the upper limb for stroke victims. This rehabilitation device includes a "serious game" intended for the transfer of knowledge that is useful to the public for the management of their own rehabilitation.

DEFINITION 4.3.– A serious game is an "educational application, whose initial intention is to combine, coherently and at the same time, serious aspects, in a non-exhaustive and non-exclusive way, teaching, learning, communication, or even information with the fun aspects of video games" [ALV 07].

4.3.2. *Organizational aspects*

Access to more data, but especially to data adapted to a specific context, data that is remotely available, leads to a number of constraints (delays, travel costs, demands associated with decision-making). The healthcare pathways, the shared responsibilities (including those of the patients) and the respect of the professional or regulatory requirements are substantially transformed. This point is illustrated in the following cases.

4.3.2.1. *The Santé Landes case*

The first case, that of Santé Landes, re-establishes the experience of the introduction of connected objects in territorial support platforms. It does this by highlighting the usage requirements at the organization level.

4.3.2.1.1. The case

This case reports the experience of connected objects in the Landes digital health project: TSN (Territoires de santé numériques – Digital Health Territories). TSN is a national program funded by *investissements d'avenir*[15] and involves five regions including New Aquitaine.

In the Aquitaine region, in anticipation of the Health System Modernization Act, a territorial support platform – PTA – has been set up. In the context of TSN, it involves implementing digital tools to enable the coordination of healthcare pathways, made available to all professionals involved in this journey, as well as a patient tool allowing both patient and

15 French investment program for innovation.

professional information to be displayed. There is an awareness that the patient must be able to better control their health and to be active in their own care.

In the TSN Landes program, there is a mobility coordination tool, a web service, available on Apple iOS and Android. The solution is distributed to almost 1,200 users. Approximately 1,300 situations have been dealt with thanks to this system since September 2015. The active file contains 800 coordinated situations.

The current objectives are to graft connected objects so that a certain amount of information can be traced back using these objects. The users of this solution are only some of the population affected by the PTA.

The terminals used at this stage of the project are iPhones and Android phones, with apps for professionals, and Androids for patients. It allows the exchange of photos, the keeping of a journal, etc. The aims of the professional tool are: follow-up and support of the patient's pathway, diffusion of information, the management of movements and orientation in healthcare services. In the area of patient monitoring, the "sign" functionality aims to present data in real time, enabling the patient's situation to be assessed at a glance. Some of this information is data from connected objects. This sign must be able to present everything that is of importance.

Below is a report on the concrete experiments of introducing two types of connected devices.

Connected glucometer

This experiment was undertaken with URPS (French regional union of health professionals) and Landes nurses. They were very motivated by the prospect of having a constant relationship with the patients through connected objects. The choice fell on a product communicating via a box in the patient's home. The homes were chosen from areas in which the broadband coverage was satisfactory. On the professional side, seven nurses were mobilized, all very keen to volunteer. The operation started in November 2016; come the end of December, three out of seven of the professionals did not want to ever hear about the device again. They have since been replaced, but we have sought to find out the reasons behind this change of attitude.

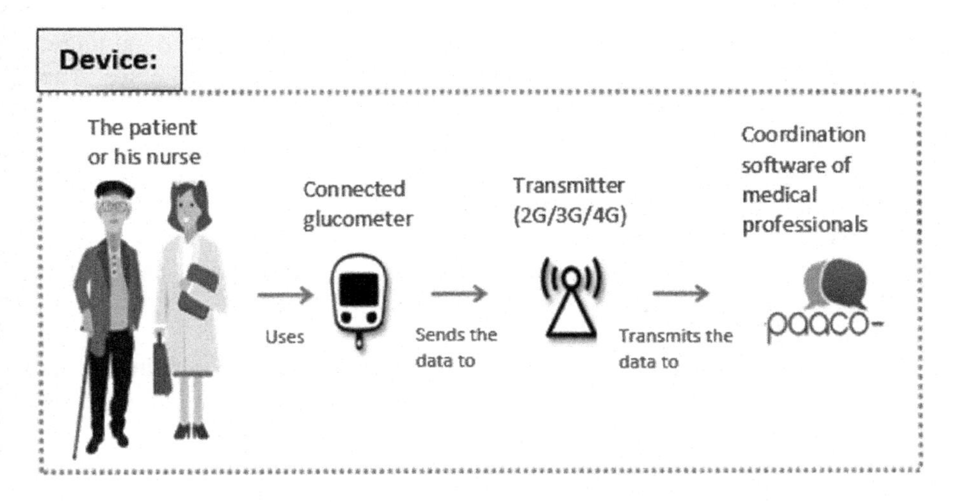

Figure 4.2. *The connected glucometer*

Their judgment is that the handling of the glucometer was complicated for the patients and that the relationship between the doctors treating the patient and the nurse was not satisfactory: the doctor did not provide sufficient support and did not pay enough attention to the requests resulting from the set-up of the glucometer. They were also not involved to a high degree.

Some obstacles were observed on the patient side. Problems with the transmission of information were noted. On the patient side, the connection was permanent, but the transmissions were carried out separately to supply the coordination tool. For this to work, the transmitter had to be constantly turned on and plugged into the mains. However, the prospect of later multiplying the number of connected objects had been abandoned: two objects were already too much.

In addition, the tool experienced some anomalies (duplications, wrong indicated dates): the adjustments to these defects were made without delay.

When there was no connection (white area) with the operator, the proposed solution was connection to the Wi-Fi box, if it existed, to overcome the difficulties in the transmission of information. However, in some places, because of the layout of the home, the Wi-Fi box did not work well. Technicians came and went several times and because of the geographical context, it typically took more than an hour's drive to get there. Sometimes there were problems with the handling of the glucometer. The platform was solicited for the resolution of this difficulty.

The multiplication of transmission equipment effectively weakened the transmission: a glucometer, a separate transmitter, a network connections – it was all too much. Motivated professionals who were called on to travel more became discouraged. The consequence was a loss of more than half of the initial nursing workforce in one month.

In summary, it appears that having two tools in the device complicates things and can demotivate the users. The proposed solution is to integrate the transmission into the glucometer to end up with a single tool.

Connected watch and remote consultation

Connected watches are part of an experiment called e-DomSanté, conducted by the Bergonié institute, a cancer center located in Bordeaux, and involving patients residing in the Landes, a territory remote from this center.

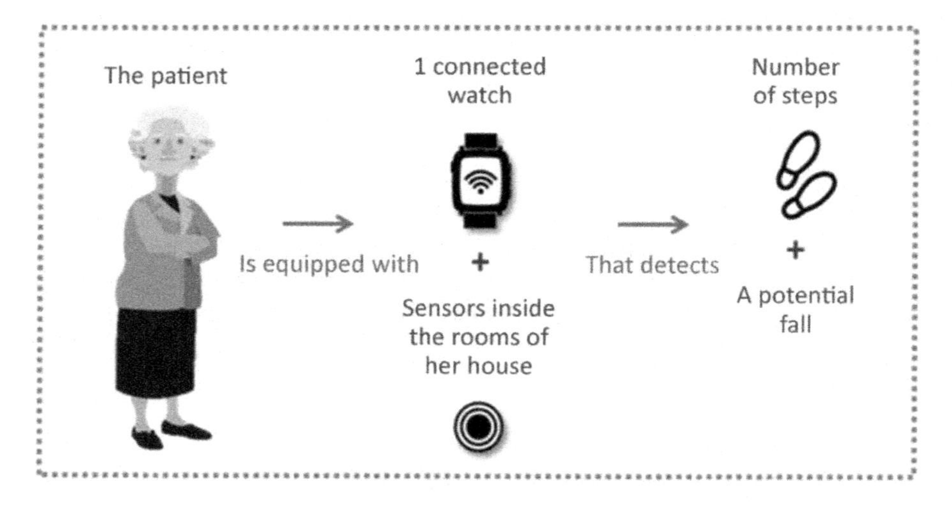

Figure 4.3. *The connected watch*

The aim was to reinforce the follow-up of individuals weakened by an advanced stage of cancer, thanks to devices that enable remote consultation (iPads) and gathering of information (connected watches). A total of 15 patients were included, the first in July 2016. They were very tired individuals who need to avoid traveling to the remote CHU.

Different devices were implemented: the tablets allowed them to ask questions, to receive visual and vocal messages and to access updated information daily. The ability to ask questions was important to help patients manage their anxiety.

The connected objects were watches worn by the patients, with sensors placed in different rooms of their house and a sensor detecting the fridge opening. The challenge was to be able to compare the declarations patients' with observations of reality. In the context of the connected watch, the device that accompanied them (a remote consultation via iPad) did not prevent the home nurse from continuing their interventions. Emergency cases were not covered.

Concerning the tablet, things went well. It was well accepted. Moreover, a certain number of people already owned one. There were a few problems regarding the questionnaire itself, but that is a different topic.

For the sensors in the rooms or on the fridge, the information is only valuable if the individual lives alone. If there is a spouse, visitors or a pet we do not know how to determine what comes from the patient and what comes from others. Therefore, the sensors also counted the number of steps of dogs above a certain size. Moreover, sometimes the sensors can fall and break. On top of this, there is the question of operating the sensors in large old houses: sometimes the signal does not pass. In addition, some individuals did not want this kind of monitoring in their home. Concerning the watches, some falls were not picked up on. In general, the experience was not satisfactory.

Facing these difficulties, it was decided to put a stop to the sensors. The watches were upgraded by making them capable of picking up footsteps, falls and bed rest.

Despite these technical problems, the patients perceived the benefit of the connected object and gave positive feedback. The possibility of remote monitoring and the ability to ask questions, is assimilated to a nursing presence.

It should be noted that the implementation of these devices was costly. To obtain a satisfactory functioning, it was often necessary for someone to come from Santé Landes or from the center of Bergonié to the individual's house, sometimes up to 3 or 4 times. During the month and a half following the initial installation, this represented up to two days per person in total.

Lessons learned

The full-scale experiment of connected objects is a novelty with which we have little perspective. In this respect, territorial support platforms, which contribute to the coordination of medical professionals around the patient, constitute a favorable environment for this type of experience. Moreover, if the case is limited to two situations, other connected objects are expected in other projects, such as a connected spirometer, for example.

4.3.2.1.2. Conclusions from the Santé Landes case

The Santé Landes case allows us to draw the following lessons:

Designing objects according to the context of intended use

It is necessary that objects synergize well with the gestures and behaviors of the individuals who use them. This question was not addressed in the program, and the devices could have been better designed. No work has been undertaken regarding the emergence of new tools that are specifically adapted to experience-based situations. The reason for this is that the TSN[16] program is under harsh time constraints, and the mobilized objects come from trade. The design would have considered the daily life into which the objects fit, drawing the entirety of the experience, clarifying the human and geographical context, the situations, putting the objects in specific situations.

Selecting the right people for experiments

It is necessary to establish some prerequisites to allow the experiment to proceed at its best. The criteria had not been worked through enough at the beginning. Motivated and driven actors are necessary and they must also be aware that the beginnings of the experiment are always slightly difficult. This includes choosing patients (for the glucometer: patients in white areas and who do not possess a Wi-Fi box could not benefit from the device). It is also necessary that the patient and the nurse have a good relationship.

16 The TSN program is an initiative set up to test digital solutions for health.

Finally, it is necessary that the doctor/nurse pairs be involved. The necessary involvement of the general practitioner was underestimated, and it is important to return to this: they are in contact with the nurse and the patient and it is necessary that they relate to the device and recognize the nurse's role in the resolution of installation problems and problems due to the complexity of such an installation. Successful implementation is based on teamwork.

Always linking the tool to the human

The connected object must be thought of as an element and not a substitute for devices which allow the patient to take responsibility in their healthcare pathway. The object cannot and must not replace the human. Doctors and nurses maintain a bond with the patient, a bond that the object can reinforce. The home-based nurses must continue their interventions. Emergency situations are not covered. The doctor and the nurse continue to have full responsibility. The connected object does not characterize an emergency and in practice, it is above all the nurse who visits the patient daily.

Supporting the first users and staying in tune with their comments

The organization also includes the support of the handling of new tools and their monitoring. This goes from initial training to support in case of problems encountered. In the context of Santé Landes, the Regional Support Officers have fulfilled this advisory role that also relays information during difficulties. The caregivers have underlined the importance that the PTA had in the smooth running of the project by being a very responsive local team. At the same time, we should not be obstinate; after a certain number of incidents, trust is no longer palpable and we must know how to give up.

Daring to rethink the starting device

In the context of the glucometer, the transmitter had many strengths (worked on 2G/3G/4G, could be connected to other connected objects for a mutualized transmission of information). In use, however, it proved to be disabling because of malfunctions and a complexification of the process. The experiment is currently being reoriented to integrate the function of transmitting information to the platform into the connected glucometer.

4.3.2.2. *Hadagio*

In the Hadagio solution, the gathered information concerning the patient allows follow-up observations that are likely to modify the work organization of the professionals:

> "Elements are sent to a module that tracks tasks and facilitates the coordination of medical professionals and social workers or other patient-involved organizations: for aid functions, actors from CCAS – Centre Communal d'Action Sociale (French Center for Social Action; various locations), etc. These actors can then organize their interventions around what happens to the patient. The information concerned is personal intervention, care and follow-up of tasks" (section 4.2.3.1).

4.3.2.3. *The Prisme case*

This case illustrates the possible impact of connected objects in the organization of work.

It is a question of developing a smart device that facilitates autonomy and the follow-up of patients when they leave their bed to reach their armchair, or vice versa. This transfer carries significant risks for patients who have motor difficulties. The solution incorporates a communicating bed embedded with sensors optimizing medical care, and an armchair equally embedded with sensors that corrects the user's trajectory thanks to a robotic module. The challenge is that the bed and armchair must communicate and understand each other. The armchair could thus position itself near the bed. Beyond the security of the transfer, we see in this example what the solution brings to professionals (risk, hardship) and an increased flexibility in the organization of work. The originality of the project stems from the integration of a new actor: Living Labs, which will benefit from funding even though it is neither academic nor an SME. This project consolidates ISAR in its vocation of LLSA. The Living Lab's contribution to the project includes the definition of the users' needs, the co-design of test benches, user assessment scenarios, the evaluation of the solution's effectiveness (service rendered) and the uses in an experimental setting and in open environments.

4.3.3. *Resilience, alert and security*

The fact of having data in good time (not necessarily "real" in the technical sense) enables us to collectively react better to deterioration of situations or unforeseeable events.

We can in this context distinguish between alerts and alarms.

Alerts are characterized by the following elements: they correspond to the crossing of a threshold: physiological parameters, levels of activity, etc. This threshold refers to an absolute value of the measured variable. The temporality is immediate. The triggered action is of the curative order, and supposes an immediate intervention.

The alarms relate to malignant trends established over a long period of time: slimming, lifestyle change, etc. The threshold of an alarm refers to a derivative of the measurement. The temporality is not immediate. The triggered action is preventive: this allows for the required time to implement corrective measures.

For example, if an individual weighs 65 kg, it is not the same if they weighed 10 kg less a week beforehand (alarm) than if their weight was identical in that previous week, regardless of the individual's physical characteristics.

The technical and organizational requirements, the associated costs and involved responsibilities differ significantly depending on the choices made. For example, in the case of Sante Landes above, the urgent cases were not covered.

On the contrary, other devices are warning devices and require specific organizational arrangements and intervention protocols: this is the case, for example, for geolocation devices that spot people wandering around when they leave a predefined security perimeter.

5

Challenges and Limitations of Data Capture versus Data Entry

Technical solutions available to patients and first-line professionals pose a number of problems, which can be different depending on whether or not the devices being used are passive or operated by the user, a distinction introduced in Chapter 4, section 4.2. The aim of this chapter is to develop this aspect. In this chapter, we will discuss the impact of these malfunctions and requirements associated with this kind of solution according to the passive or active nature of the interface with the human: clinical questions associated with the exploitation of new signals from the human body which are compatible with the patient's everyday life, requirements of remote operation and facets of devices used in an uncontrolled environment.

5.1. Uses confronted by technique

In this line of work, most identified cases undergo problems due to the confrontation in the field between patients and other users, which are caused by the technical reality of the connection. These concern a variety of aspects: the inadequacy of the solution in the given context, difficulties implementing control and using the solution, up to and including functional and technical problems.

This point is illustrated through feedback from the company BePatient in the field of "research" (see Chapter 2) and the case of Santé Landes.

Chapter written by Norbert NOURY and Robert PICARD, with contributions by Marie-Noëlle BILLEBOT, Frédéric DURAND-SALMON, Myriam LEWKOWICZ and Henri NOAT.

5.1.1. *BePatient: feedback*

In one of its projects, BePatient aimed to collect everyday data for research and found that after a few weeks, the pace of including new patients was slow, tending towards zero after a few months. Early evaluation of the benefits from the patient's point of view was performed with a strong focus on the use of connected objects. Thus, it has been shown that this use was a source of discontent. Many requests received by the hotline were about these objects, problems of use and updates.

More precisely, the results of this evaluation are the following.

Regarding the recovery of data and objects, only 44% of included patients took at least one measurement a week, without any obvious link to the patient's age. Among these reasons, one concern is the lack of application updates linked to connected objects; when the IOS of some of the delivered tablets were automatically updated, the connected object was no longer usable.

Concerning the objects themselves, in fact, two out of three objects were well used: the scale and blood pressure monitor (used for at least one measurement per week on average). However, the activity tracker proved to be extremely problematic.

Regarding the provision of learning and messaging functions, they went well. Patient involvement in the study was good, with several visits a week and an average visit duration of 11 minutes. The most-visited page was the dashboard, which shows the patients' appetite for monitoring via connected objects.

Following this usage study, several measures are taken: stop providing defective connected devices; improve ergonomics (with single application for measurements and for other questionnaire functions, learning, etc.).

Other causes are being researched to improve inclusion: simplification of the process and co-study training (clinical research associates). In fact, this was the most important thing; it was necessary to interest prescribers and improve their own interface. This dimension had not been sufficiently thought through.

5.1.2. *Santé Landes*

The Santé Landes case (section 4.3.2.1) reveals similar difficulties:

> "The multiplication of transmission equipment effectively weakens transmission: a glucometer, a separate transmitter, a network connection, is all too much. Initially high motivated professionals, called on to travel more often than not can become unmotivated. The consequence of this is a loss of half of the initial nursing workforce in one month".

Therefore, poorly controlled use and the lack of *ex-ante* evaluation of connected objects can jeopardize an entire project or solution, even when such objects do not constitute a decisive element in the design of the overall system.

5.2. Different sensors for different uses

5.2.1. *Biological measurement: challenges and constraints*

Biological measurements carry many substantial issues. For example, a research laboratory represented in the group was interested in integrating electrodes into bandages in order to monitor wounds. Indeed, at the level of wounds, there is no skin and the resistance of the medium is weak. In healthy areas, resistance is higher. We therefore have measurements to characterize the tissue. But even better, we can also produce an electric field in the bandage to accelerate wound healing.

Measurements taken by ambulatory sensors represent tremendous potential: measuring the stress levels of blind people while moving; driver vigilance depending on context, from the human–machine interactions; monitoring daily activity using intelligent environmental sensors: circadian rhythm and risk situations, etc.

In general, the collection of physiological information is not only a sensor problem, but also there are also issues with the interpretation of data, the training of clinicians and the number of informed clinicians (see Example 5.1). It would be useless to send information if there is no one to interpret it.

Transversal reflection must be carried out, which goes from tissues (physical models) to biosignals and transduction mechanisms, which will result in the design and evolution of sensors.

EXAMPLE 5.1:– An ECG sensor patch would allow ECG data to be taken by a simple patch placed on the subject's chest. However, the signals delivered by the device are not of the same type as those sent by conventional ECG devices whose data capture occurs at the extremities of the limbs. Such a device requires training clinicians in a new type of interpretation.

Thus, it is not enough that technical instruments exist; it is necessary and important to provide lessons, to monitor the situation in an ambulatory fashion, to observe, to collect information on what is happening in the context in uncontrolled situations, and finally to analyze and interpret it. For example, it is useful to associate information allowing the identification and tracking of patients by patches or other embedded devices.

5.2.1.1. *An opportunity to expand our knowledge*

An example of a physiological function which can be measured by different biosignals is heart rate: auscultation, pulse, electrocardiogram, PPG, the Doppler effect (an antenna sends a radio signal according to the cardiac pulse depending on the surface of the body) and cameras (from a video, we can extract useful information: the frequency of variation of tissue opacity is a function of the frequency of oxygenation of blood by the cardiac pump). The choice of measurement system depends on the context: is the person stationary or mobile? Is it an establishment or a home context (in an uncontrolled environment)? The goal is to take measurements which can be compared to a "gold standard", or measurements taken from the same patient at another time (the patient is then their own reference: p-Health).

These measurements represent a potential for new discoveries about human physiology thanks to the ability to search through an immense database of longitudinal information collected at home by large cohorts: we can look at an entire lifetime, or at the population of an entire region. Medicine only has a partial vision of phenomena: we know the circadian rhythm well enough, and we have some ideas about seasonal rhythms. But for variations over a year, we know nothing. By working on a large regional cohort, for example, it will be possible to analyze geographical and environmental impacts regarding this data.

To return to the topic of life and age, we know that autonomy is acquired at the beginning of a lifetime and is lost at the end of a lifetime, after a period of decline in capacities. This loss is variable depending on the individual, and can occur at older or younger ages. Measurements can be established to detect decreases in autonomy, and actions can be made in order to allow the individual to live at home for longer, using adapted aids and preserving the balance of autonomy.

The question now is what can be measured and where?

5.2.1.2. *Skin, a prime measurement site*

Skin is a surface that covers the entire body (1.5–2.3 m^2). It can show information regarding vital organs and peripheral physiological functions (Figure 5.1).

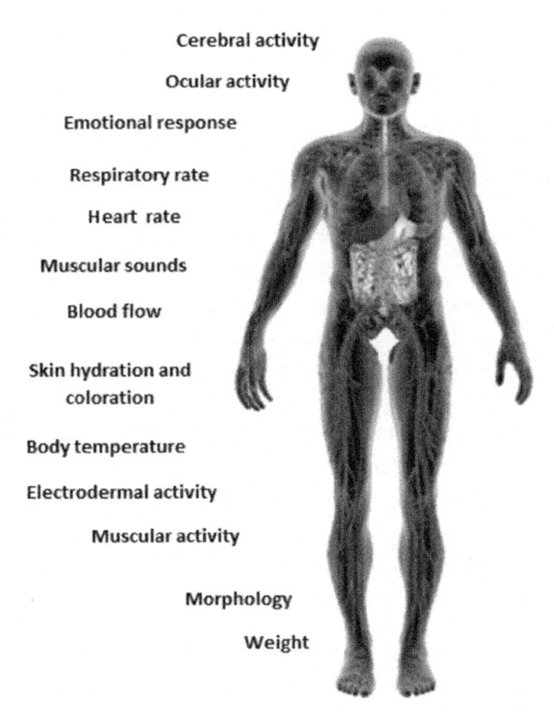

Figure 5.1. *The skin as a measurement site (source: INL – CNRS)*

The sensors placed on skin are minimally invasive, and induce few lifestyle changes. Skin is therefore a favored operating surface.

The arrival of textiles carrying sensors is a new concept; however, several difficulties remain to be solved. Thus, regarding the power supply of embedded technology, batteries are not yet small enough to be comfortably integrated into clothing. Furthermore, batteries and materials are not compatible with the usual practices of laundry care (washing, drying).

5.2.1.3. *Time measurement*

How much time can patients dedicate to data capture? Measurement practices tend to decline over time.

We refer to "real time" which consists of a time compatible with the requirements of medical time. We can recall the distinction previously proposed between "alerts" (requiring immediate intervention) and "alarms" (associated with a long-term issue).

Time measurement must also include difficulty for the patient: an annoying or stigmatizing measuring device imposed over a short duration.

Time measurement is thus a compromise between the length desired by the health professional, the time bearable by the patient depending on the device used, and for long-term measures, the ability of the patient to make an expected contribution. These requirements may also have an effect on the choice of sensor or application.

The reliability requirements are of course decisive, and are more so when the measurement is frequent and performed over a long term (problem raised in the Do Well B. case, for example – this case is presented in Chapter 6).

This question will become increasingly important as medicine evolves from curing critical situations to long-term prevention and monitoring (see Chapter 11).

5.2.2. *Sensors and behaviors*

Although wearable devices have the potential to facilitate changes in health-related behaviors, these devices alone do not necessarily lead to changes. In fact, the successful use and potential health benefits linked to these devices depends more on the design of patient engagement strategies than on technological specifications. These engagement strategies are combinations of individual encouragement, social competition and collaboration, and effective feedback loops which can modify human behavior.

This topic joins the debate on open observance in Chapter 2:

> "Concerning chronic diseases, observance is expected throughout. The statistics clearly show that this is a public health issue. But what is the cause? How and under what conditions would being connected improve this observance?"

5.2.3. *Environmental sensors*

Environmental sensors are connected objects capable of providing various types of information: location, position, the individual's movements and contextual elements which can be compared to data collected via sensors embedded on or implanted in the individual, including the validation of alarms, like in the case of falls. They pose particular ethical problems as they are a type of surveillance which can affect the individual's private life, even their privacy, depending on where they are placed. This point is particularly sensitive in the case of video capture. This type of sensor can also be associated with robotic devices with which they interact to allow them to adapt to the context or the need of the person with whom they are meant to interact. The monitoring of the individual and their health is not the first objective for some sensors as air quality sensors, light sensors, smoke detectors, etc. exist. However, data they collect can, by cross-referencing with data from other sources, contribute to the production of potentially personalized health information, and eventually the generation of alarms.

5.3. Applications and actuators

5.3.1. *The question of meaning for the patient-user*

Equipment worn by an individual, or integrated in their living space, concerns that individual, whether it is expected of them to interact with it or not. In the second case, as for all domestic equipment, the individual intervenes at the time of its installation. This cannot be undertaken without the individual's agreement, which can be costly or can disrupt their daily life. This is also the case when the equipment is faulty, of which the individual is necessarily informed. Their behavior is the subject of conjecture by the designer. The attitude and behavior of the individual can be key, for example concerning the power supply, depending directly on the meaning they give to that object, to the service that it brings them and to the effort that their eventual contribution demands, if any. This dimension is naturally all the more essential as the interaction is frequent, sometimes daily, or even occurring several times a day (a glucometer used by a diabetic individual, for example). The arduousness of the interaction must be compensated by a perceived benefit where the individual is aware that a risk was avoided.

5.3.2. *Integration into the ecosystem*

A technical device integrated into an individual's daily life must take into account the fact that they are a social being. They are never entirely alone: they can live as a couple, receive daily visits from carers, relatives or neighbors, whether they are linked to the individual's health status or not. These visitors can disrupt the operation of a system (example of the sensor detecting the opening of fridge doors, etc.). Voluntary use, at times chosen by the individual, of a measuring device or a data entry system may constitute a social constraint: the necessity to remove oneself from an ongoing relationship or a group (glucometer), to prohibit oneself from traveling, partaking in leisure activities, at least at certain times or for certain durations, or to be perceived as stigmatizing and thus remain hidden (breathing apparatus not compatible with family accommodation).

Among the connected objects presented in the Santé Landes case (section 4.3.2.1), a watch worn by the patient, sensors placed in different rooms of their home and a sensor detecting the fridge door opening were implemented to compare patient declarations to real observations. The reality is that when the patient does not live alone, when their house is too big or too old, these devices do not work. They have to be:

> "...drawing the entirety of the experience, clarifying the human and geographical context, the situations, putting the objects in specific situations"[1] (section 4.3.2.1.2).

These aspects can be anticipated by going through a Living Lab.

1 Santé Landes.

5.3.3. *Motivation and domestication*

It is necessary to question the demands induced by the presence or the use of a device during the duration of its occupancy, as well as the duration during which it is supposed to perform the function that was assigned to it. Classic ergonomic analysis uses the term "acceptability". This term is associated with a passive attitude that is adopted but not necessarily relevant (see Part 1). Not only can the individual be proactive, be an actor of their health, but this proactivity can be a condition of success of the solution. This is all the more important as the duration is long and/or the discomfort suffered is significant (hence, the possibility of renunciation of precise but unpleasant measures in the case of a long-term follow-up; see Chapter 4). Many studies show that acceptability is also due to design[2], and yet, in most cases, only medical experts are selected for the design of devices.

The confrontation of an individual (or a possibly heterogeneous group) with a new object can lead to an immediate rejection. If the desire to try is present, the first attempt can put a stop to this initial wish. This desire can dissipate over a short period of time (the findings of Diabète LAB are that all new objects destined for diabetic individuals are no longer used after a maximum of a year). Mobile applications regarding health have a usage time after download that rarely exceeds a few months. The term "appropriation" signifies that this problem is solved. The user "endorses" the object and implements its usage into their routine. The term "domestication" indicates a superior degree of sustainability of use: one where the object "is part of" the home (*Domus*) or an environment which is familiar [HYN 09].

> "Domestication is practice, it involves human agency, it requires effort and culture and it leaves nothing as it is" [SIL 05].

Ex-post evaluation, in real life, ideally several months or even years after the first use, only allows us to note it. This reality directly conditions the impact of a solution, including in terms of public health. It should be part of ex-post clinical evaluation, in the same way as the monitoring of eventual secondary effects.

---

2 The question of the role and modalities of design are developed in Chapter 7.

5.3.4. *Self-measurements versus health measurements*

We have previously mentioned the question of measurement: its foundations, its ambitions or the potential presence of metrics. This question, which highlights the importance of the purpose of measurements in order to characterize them, takes on a particular importance in the field of healthcare.

Measuring for who? For me, in order to know where I am in terms of my health status? Compared to myself during previous measurements? Compared to a community in which I recognize myself and which serves as a reference, or even as a support? Compared to a sanitary or social ideal, collective or corresponding to my identity? To measure for others, for whom I take care of or who demand it "for my own good"? It is considered that the standard reference can be extremely variable: myself, the community in which I recognize myself; the "cohort" in the medical sense in which I am included or with whom I am assimilated; the people of my age range; the general population (the "average citizen") and so on. Individual or community measurements can be satisfied with qualitative data, provided that there is a pre-existing trust: trust in oneself, trust in the other with whom we identify and with whom we are connected. However, in other cases, metrics remain ideal, or failing that, a recognized standard: a scale of values, categories that are recognized by the scientific community as relevant, that are published in international community journals; a guarantee of scientificity, objectivity and seriousness.

Measuring is therefore comparing one's own observations with those of a qualified community, according to indisputable metrics, possibly very different from common-sense thoughts. There can thus be an inconsistency between the effect on the individual of their self-measurement and the interpretation of those to whom the measurements will be transmitted[3]. This inconsistency can ruin the effects of an "objectification" supposedly achieved by measurement. Self-measurement for one person is subjective and self-measurement for others will be worthwhile according to what the other expects and the image that is created by the norm.

These dialectics deserve to be elucidated upstream, before the choice of data or information to collect in order to monitor a state of health. In the Hadagio solution, for example (section 4.2.3), measurements generated by

---

3 See section 5.4.

sensors are identified as such and they are clearly differentiated from the measurements made by a medical professional (measurement of blood pressure by a nurse versus self-measurement performed by a sensor). These measurements do not have the same value.

It is conceivable that in the future, self-measurement standards can be established for certain parameters. The question of measurement reliability for medical use is a topic to which we will return in Chapter 10. However, subjective reference may prevail in other measures (diversity and quality of social links, comfort of housing, for example), as is the case with the "Barometer" tool. In all cases, it is essential that the user identifies changes in the values shown by self-measurements and is in a position to input them and report them if required.

5.4. Value of the data

So far, we have focused on the value of data in the sense of the interest they potentially represent for the user and the medical team. In the following section, we develop on the conditions of producing this data so that they are indeed as valuable as expected, for the different potential users.

5.4.1. *Data qualification*

The data obtained from connected devices have no intrinsic value: this value depends on both the conditions in which they were obtained as well as the skills and knowledge of the person who qualifies (and potentially filters) and exploits them.

Therefore, for example, in the Hadagio system (Chapter 4):

> "The value of the data depends on the way in which they are collected: if it is done by a medical professional, it will be validated medical data; if it is done by a frail individual or a carer, it will be indicative data" (section 4.2.3.1).

This qualification involves the very design of the "connected" data collection device, and the requirements set at this stage on the conditions for inputting or capturing of this data. In this respect, Living Labs have a role to play, in that they propose experimental conditions that are as close as possible to the real environment of use.

For triggering alerts, a decision process is configured by the doctor treating the patient for each sensor:

> "The interpretation of the value of each sensor goes through a rules engine. A simple rule means that only one condition triggers the alert. A multiple rule means that multiple conditions can trigger the alert – the number of conditions is not limited. A compound rule is a rule where context, history, calculated values, etc. are all considered" (section 4.2.3.1).

5.4.2. *Compared values of captured data versus collected data*

The question may arise as to why some solutions prioritize the transmission of information collected by a patient while some information, such as weight or temperature, could be directly transmitted from a connected object.

In the MoovCare system presented in detail in Chapter 6, patients must input into a questionnaire various pieces of information about their symptoms including measurements, without the use of connected objects, although this is envisaged.

The elements that lead to this choice for MoovCare are worth mentioning. These are the following:

1. The patient already disposes of – or can easily find at their pharmacist – non-connected objects that enable the necessary measurements. The fact that they are not extremely precise is not as important as their capacity to measure variations (fidelity).

2. There exist few connected objects that are reliable and economically affordable: in any case, the project holders did not find any that met their requirements.

3. As has already been mentioned, connected objects impose a certain amount of logistics and an increased complexity of solution updates.

4. The formats and standards of remote transmission are a constraint, especially if you do not want to go through GAFA[4].

5. Finally, there is the issue of trust of practitioners. These connected objects are recent and poorly known, which is not the case for devices currently used by patients.

Conversely, inclusion in the solution of connected objects, of a patient's material component, can be justified by other considerations. This is the case for Hadagio, presented in Chapter 4:

> "In the exchanges that took place at the very beginning of the design of this solution with hospitals in Toulouse, Marseille and Bastia, practitioners have expressed the need to supplement the declarative with factual elements that allow them to better qualify cases justifying the need for interventions, and possibly doing so upstream: with more precise elements and more frequently. This makes it possible to control the declarative information by frequent and regular monitoring, to not be dependent on the declarative for the coordination of interventions. In fact, it is not a question of controlling what happens, but instead of knowing about it"[5].

4 GAFA is an acronym designating the four American giants of fixed and mobile Internet: Google, Apple, Facebook and Amazon.

5 Blot N., Noat H. (SESIN), Presentation, Forum LLSA working group, Paris, October 27th, 2016.

6

Models and Algorithms

Information and data from connected health devices are not only used by the patient, who is aware of health parameters, but also by the health care team monitoring them. According to the model presented in Chapter 1 of this book, they are also intended to be exploited in large numbers to develop new knowledge and make decisions.

The report by the Member of Parliament for the Essonne department, Cédric Villani, a mathematician who won the Fields medal on 28 March 2018 for his work on artificial intelligence (AI) [VIL 18], identifies the healthcare sector as particularly promising:

> "Artificial intelligence in healthcare opens up very promising prospects in terms of improving the quality of care for the benefit of the patient and reducing costs – through more personalized and predictive management – but also their safety – thanks to reinforced support on medical decision and better traceability. It can also improve citizens' access to care, thanks to pre-diagnostic medical devices or guidance concerning the healthcare pathway" [VIL 18, p. 196].

The importance of medical data, the benefits of algorithms and models which underpin them to further our medical knowledge is also recognized by physicians. The white paper published in January 2018 by the CNOM[1]

Chapter written by Pierre BERTRAND, Daniel ISRAËL and Robert PICARD.

1 French Medical Board.

[CNO 18] on artificial intelligence, data and algorithms expresses the following opinion:

> "Algorithms and artificial intelligence will be our allies, used as an essential contributor in therapeutic decisions and strategies, as well as medical research".

In this chapter, we will develop some elements to understand these notions and associated challenges by illustrating them using concrete cases within the framework of collective work.

6.1. Representations and algorithms[2]

The technologies which underpin what we have called the "representation" in the reference model (Chapter 1) are diverse: computers, databases allowing links to be established, applications allowing them to be manipulated, to be organized, to be counted and to count statistical correlations.

Recent times are characterized by a considerable increase in computing power and the amount of data collected by devices. New knowledge can thus be obtained. Concepts of "Big Data", artificial intelligence including "machine learning", are the subjects of infatuation and speculation.

We will return to the topic of these technological constraints in section 6.3. Here, we limit ourselves to illustrating the value of algorithms in the context of connected healthcare.

DEFINITION 6.1.– **Algorithm.** The Larousse[3] dictionary defines an algorithm as "a set of operating rules whose application makes it possible to solve a problem, stated by using a finite number of operations. An algorithm can be translated, thanks to a programming language, into a program which is executable by a computer".

---

2 This section is loosely based on remarks by D. Cardon, sociologist, director of Medialab at Sciences Po, during his speech at the seminar of the Conseil Général de l'économie, in September 2017.

3 https://www.larousse.fr/dictionnaires/francais/algorithme/2238#ELmCo7Ic2dT1o5L0.99.

The concept of "problem solving" is therefore an integral part of the definition of an algorithm, and we recognize that it is at the heart of value creation. But in practice, it is conditioned by a number of the characteristics of the system in which it is placed, particularly beyond the technical limits, the relevance and the adequacy of data which is presented to it. Certain algorithms are designed to evolve in function of the statistical distribution of data, which itself varies over time. In favorable cases, they are then able to predict more and more reliable results when the volume of data grows: this is commonly called machine "learning". The value of the algorithmic system can therefore increase over time if the volume of data has certain qualitative characteristics and becomes sufficiently large.

The issues surrounding algorithms are numerous and are of economic, social and legal nature. From an economic point of view, it can lead, by the rupture it creates in the ecosystem, to dominant positions and to a disintermediation. The arrival of connected objects must be accompanied by requirements of interoperability and standardization to limit these risks.

Social issues are more complicated to deal with: we must ensure non-discrimination, which can result from the use of certain media, cultural or political characteristics. It transforms the working perspective, through tasks undertaken by software robots. The role played by learning, not to be confused with machine learning and the machines' capacity to integrate environmental data to act effectively, must be reconsidered.

Regarding legal issues, this consists of checking the algorithm's conformity to the rules of law, to put the human and machine each in their place. This calls for the state's ability to play its regulatory role.

What are the expectations of algorithms? Neutrality regarding problems to solve is not possible: the owner introduces bias which creates value for them; transparency: this is not always possible, as this makes the algorithm more vulnerable, especially to hackers; loyalty: that is to say what one does and doing what one says; finally, auditability: which means we can explain the algorithm's actions by getting out auditable parts.

Whether it consists of problem-solving or predictive capacity, algorithms carry a significant and potentially growing share of knowledge resulting from data collected through connected health. It is also necessary for this knowledge, which is sometimes counter-intuitive, to be recognized by practitioners and affected patients, and integrated into their practice. Their diffusion and acceptance by the social body depends on many factors, including the diversity of interpretive capacities in different categories of actors. These aspects are illustrated in Chapter 7.

6.2. Artificial intelligence in health[4]

Historically, explanatory models of the functioning and dysfunctions of the human body were developed empirically, based on a few observations. Nowadays, models used in modern medicine use statistics and data, which quantifies their relevance in a large number of cases.

However, it is possible to take advantage of the digital world. Mathematics will contribute by being combined with more traditional models. Classical modeling will be supplemented by information coming from equations. Thus, we can fit a generic model to a singular case.

The term "learning" is sometimes used. Realistically, it often consists of statistical correlations, possibly counter-intuitive ones. The results of these observations allow us to make decisions based on statistics, even if we do not have an explanatory model. In addition, these results can lead to the formulation of hypotheses which may lead to methods different from those used in the past. Models can mobilize differential equations to tackle cancer, pathologies of the heart, of the brain and so on.

For example, the mathematician Thierry Colin used equations to monitor the evolution of certain forms of cancer on a daily basis, and predicted this evolution, which has the consequence of being able to limit chemotherapy to only be used during periods of risk. It also allows us to choose the right moment to perform an operation. It does not only involve statistics, but also prediction and phenomenology.

The combination of classical and mathematical approaches represents a big challenge, with possible generational conflicts between professionals. Significant efforts in the training of professionals are to be planned for.

---

4 This section is inspired by Cédric Villani's intervention at the 2017 G5 annual seminar.

6.3. Issues and limitations of algorithms in health

6.3.1. *Issues*

Mathematical models, embedded in algorithms, can lead the latter to be more or less fast and greedy in memory and power usage.

For example, the FFT (Fast Fourier Transform) algorithm allows a much faster calculation of Fourier transform, but it is still the same model. A different mathematical modeling technique was introduced by the "wavelet transform" representation, with various algorithms.

Algorithms sometimes carry an enrichment of models and therefore potentially new knowledge which can be mobilized to make medical decisions. The position of physicians is that if they want to conserve a future role in the organization of care, then it is essential that they control the creation of algorithms; otherwise, they may become nurses serving commercial societies or insurers who will tell them what to do, when and how. Faced with this position, the industry defends the idea that the algorithm is an industrial secret which lies at the heart of effectiveness and competitiveness of solutions, which are becoming more numerous in terms of use.

6.3.2. *Utility*

The diagnostic or predictive value of an algorithm is revealed only after a large number of tests during which data accumulates. Hypotheses are made and challenged, for others which may be counter-intuitive, not consistent with practitioners' prior opinions. The utility of an algorithm is all the more difficult to prove that it does not possibly have an explanatory model capable of rationally justifying the result. Reciprocally, mass tests can show the falsity of clinical institutions.

The MoovCare case (follow up of lung cancer) is presented in section 6.4.1. Below is an example of an algorithm whose utility was able to be demonstrated. Patients who used this application and followed advice presented a one year survival rate of 79% as opposed to 45% for the control group.

6.3.3. *Optimization*

We brought up the need for long series of observations to refine an algorithm and make it reliable. This reality continues beyond the decision to launch clinical trials. During the testing phase, it is methodologically impossible to modify the algorithm. The protocol can, for example, join a positive result provided by an algorithm with the obligation of a physician's assessment confirming the judgment on the state of health, with the possible consequences in terms of treatment. Refining the algorithm then modifies the decision system and afterwards distorts results with respect to the initial reference.

6.3.4. *Limitations*

The effectiveness of an algorithm is measured primarily by its computation time, its RAM consumption (assuming that each instruction has a constant run-time), by the precision of obtained results (e.g. with the use of probabilistic methods, like the Monte Carlo method), its scalability (its ability to be effectively parallelized), etc. The computers on which these algorithms run are not infinitely fast: the available computing time on a machine (computer, "smartphone" or "smart device") remains a limited resource despite a constant increase in performance. An algorithm is then said to perform well if it uses resources available to it sparingly, namely CPU time, RAM and power consumption, which is linearly correlated. The analysis of algorithmic complexity makes it possible to predict the evolution of computation time needed to bring the algorithm to completion, according to the quantity of data to be processed.

6.3.5. *Input constraints*

The inputs correspond to data from sensors, shareholders and applications. In an ideal world, all information which could constitute a signal, even a weak one, of the evolution of the health status should be taken into account during research into a predictive or diagnostic algorithm. In fact, the hardship and the constraints associated with these devices do not allow their use at all times, whatever the scientific or clinical interest of such measurements. Therefore, if the predictive or diagnostic capability of an algorithm depends on its capacity to mix the most relevant observations in correct quantities to draw significant correlations, this collection is itself constrained by the acceptability of the measure which can be painful, stigmatizing and disabling for various reasons. The field of the observable universe is accordingly reduced.

The dependence of inputs is not a dependence on the sensors. In the case of the ANR Do Well B. project (see section 6.4.2), it is estimated that in 1 or 2 years, we will have cheap, good quality sensors. In addition, as the presented algorithms are not "sensor-dependent" (in the way that they use the maximum resolution possible, i.e. the measurement of each heartbeat), at that moment it will be possible to shift from academic research to mainstream applications.

6.3.6. *Power*

An algorithm works on characteristic data, contributing more or less significantly to the assessment of a monitored individual's health status. The amount of data necessary is potentially high: significant data can sometimes be calculated or filtered at the collection site to reduce the time and cost of processing. But these operations at the source also require a local power source at the sensor level, which is associated with a certain amount of electricity consumption and autonomy of the equipment if it is not permanently connected to a power source (in the case of embedded or mobile equipment). Power consumption of a smart sensor ("smart device") is proportional to the number of operations performed, namely calculation time, and therefore ultimately plays on the size of the batteries or on-board batteries. We therefore have a dual interest in having a fast algorithm to process data in real time and not consume too much power.

6.4. Case studies

The two cases developed below illustrate the contribution of algorithms regarding health, both in research (Do Well B.), and in terms of proven clinical results (MoovCare).

Since algorithms evolve as experience accumulates, it becomes very difficult to have conclusive and scientifically-based results. As a matter of fact, the algorithms are stabilized during the studies. However, it happens that sometimes, as in case 2 (MoovCare), the questions to be answered by patients may be modified, improved and validated. For MoovCare, they were tested on a large number of patients.

Another topic to be discussed is the time required by CNIL (the French Data Protection Authority) for the procedure. If this time is too long, it would be a real penalty for the development of algorithms by French doctors.

These aspects are the subject of particular attention in the Villani report, which argues for the implementation of "innovation sandboxes" with a "temporary alleviation of certain regulatory constraints" allowing an "iterative design" and "real time experiments closer to the user" [VIL 18, p. 199].

6.4.1. *Clinically-validated algorithms for early detection: MoovCare*

The following case, "MoovCare", illustrates how an algorithmic approach can lead to a significant improvement in care, both economically and in terms of quality of life and the medical services provided. The clinical evaluation of this solution is presented in detail.

6.4.1.1. *Foundations of the clinical approach and conducted trials*

The concept is based on the fact that most recurrences, in the case of lung cancer, are symptomatic (80 to 90% of cases) and that most of the time the patient suffering from symptoms may wait several weeks before booking a consultation which can delay the diagnosis of relapses. This delay results both from the reluctance of the patient, who is finishing a very intense treatment, to find themselves once again in this situation, and also sometimes because of administrative issues or even due to questioning the legitimacy of contacting their oncologist. However, an increasing number of patients are connected: weekly clinical reports by electronic forms associated with algorithmic analysis have demonstrated in clinical studies, on a model similar to those evaluating drugs, many clinical benefits for the patients.

The solution is based on a questioning of 12 symptoms for oncology follow-ups.

6.4.1.2. *Operation*

After verifying patient eligibility, the oncologist prescribes a follow-up by MoovCare. MoovCare, being a web application, does not need to be downloaded via online stores and can be used on any platform with an Internet connection. The application is prescribed by the oncologist. The oncologist ensures that the patient is eligible for a follow-up by the application.

The patient fills in, once a week, a questionnaire of 12 pieces of clinical information that are instantly analyzed by MoovCare algorithms. This transmission is secured by an encryption in accordance with ANSSI recommendations. The server receiving the information complies with French Health Data hosting regulations and European requirements (the regulatory compliance differs from country to country and it is always ensured).

The algorithms analyze the association and the dynamics of the symptoms, indeed, the answer to a single question does not trigger the alert alone (for example, it would be necessary to simultaneously observe an overweight person, a depressive situation, etc. for a trigger). This alert is in the form of an electronic message and/or SMS addressed to the medical professional. The professional has personal access to the application and a dedicated space, which allows them to monitor the patients whom they have registered. They are able to view all of their history and thus have a focus on the alerts.

6.4.1.3. *Method*

At first, the clinical studies were conducted by Dr Denis, including phase III, funded by Sivan Innovation, whose intermediate results were presented at ASCO 2016 and published in the JNCI in 2017 and of which the final results will be presented at ASCO 2018. It then consists of the design by and for oncologists, in a collegial approach. This approach enables the provision of a high level of evidence, since it takes into account, from the early phases of development, the vision and medical requirements. By acquiring the product of this design, the Sivan Innovation company, whose founder is Daniel Israël, has placed itself in a position of co-design with these practitioners.

Concerning the method:

– Groups of experts in the domain have the role in defining the key symptoms representative of relapses or expressed complications and the combination rules that have been validated during the phase II trials.

– The integration of patient feedback is achieved over time: compatibility of access with a smartphone, different brands of mobile phones, taking into account the demands of relatives and carers, improved wording of questions and so on.

– Similarly, practitioners' feedback is integrated, concerning for example, the ease of connection, the transmission of alerts, their number, their relevant frequency and so on.

6.4.2. *"Do Well B.": towards an objective measure of stress*

The second case concerns the introduction of permanent monitoring devices of physiological parameters in order to prevent problematic behaviors. It is by developing an algorithm that the link between these two aspects can be realized. This type of device obviously raises an ethical question, a dimension that will be developed in Chapter 9.

However, at this stage, there is not a reliable sensor providing continuous measurements of physiological signals, those that are reliable enough to develop an application for the general public.

6.4.2.1. *Introduction: measuring anxiety in autistic individuals*

People with autism spectrum disorder often have a different sensory integration compared to non-autistic individuals: they do not perceive or feel the same things. Therefore, events that we consider totally harmless can become a source of anxiety (a flickering neon light, vacuuming sounds) leading to an emotional overload that may not be externally visible; the autistic individual may seem calm and relaxed, even though they are not, which can lead to emotional behaviors (anger, aggression). In addition, many autistic individuals may have difficulties verbalizing what they feel and fail to spontaneously escape a stressful environment.

Detecting in real time an internal stress invisible from the outside.

The aim of the Agence National de Recherche (ANR, French National Research Agency) "Do Well B." program was to identify, thanks to algorithms of statistical analysis of physiological signals, in real time and in everyday situations, increases in stress levels, in order to send an alert either to the individual, or to their relatives or carers, in order to curb the rise of anxiety and avoid problematic behavior. These algorithms could also be used in other populations with anxiety disorders that affect their well-being (people with anxiety, elderly people, etc.).

6.4.2.2. *Long-term goals*

A long-term goal that has not yet been reached is to replace the static self-testing of stress with numerically quantified stress measurements and to calculate their variations over time. Applications may concern mobile health (m-health), activity following medicine, autism (Autism Spectrum Disorder), etc.

Today, stress and mood are measured by salivary hydrocortisone, pupil dilation or self-quantification, performed hourly. In the future, we could continuously measure stress and mood levels, calculated in real time, by the use of physiological signals measured continuously by non-invasive sensors and algorithms implemented in smartphones.

6.4.2.3. *New goals, new data*

It is becoming possible to monitor physiological signals outside of hospitals with new devices. These devices enable the collection of time series of physiological signals in real-life situations, with non-invasive and easy-to-wear sensors (bracelets, watches, etc.). Being lightweight (typically 50 g), they enable a Bluetooth transmission of data such as electrodermal activity (EDA), blood pressure (BP), wrist movements (accelerometer), heartbeat (HR) and skin temperature on the wrist (that is not 37 °C but ranges from 32 °C to 35 °C). It is possible, for example, to identify periods of nocturnal electrodermal activity corresponding to dreams.

The monitoring of physiological signals in everyday life opens up new perspectives. Long physiological series bring in information about the Autonomic Nervous System (ANS), which by definition is automatic, meaning that it escapes our consciousness.

Physiological signal series can be retrieved over a long period of time: 50,000 heartbeats for one night, for example.

The temporal complexity of the algorithm corresponds to the amount of electricity consumed by the battery. We develop algorithms whose consumption varies linearly with respect to the N quantity of processed data.

6.4.2.4. *The autonomic nervous system*

The autonomic nervous system is composed of two branches: one is the "fight or flight", which is a sympathetic nervous system. It dilates pupils, accelerates heart rate, puts a break on digestion and loads glucose into the blood. These elements allow humans to be reactive when confronted with danger (fight or flight… when our ancestors saw a mammoth or a tiger, for example). The other component, "rest and digest" is the vagal system or parasympathetic system. It acts in the opposite direction: narrowing the pupils, slowing down heartbeat, slowing down breathing, but increases the activity of the stomach and the intestine.

These two antagonistic systems have an impact on physiological signals:

The sympathetic system increases electrodermal EDA activity, pulse and heart rate variability[5] and sinus variability in low frequencies (0.04 Hz, 0.5 Hz).

The parasympathetic system slows down the pulse and the sinus variability at high frequencies (0.15 Hz, 0.5 Hz).

Cardiac activity is an indicator of the proper functioning of the autonomic nervous system. The domains of application are in cardiology, physiology and psychology (well-being). The analysis depends on the duration of observation, the domain of observed frequencies and the use of nonlinear methods.

As previously mentioned, the used sensors allow real-life measurements in a minimally invasive fashion. Heart rate is a much more interesting measure than salivary hydrocortisone levels or pupil dilation that is not easy to observe continuously. Within 24 hours of heart rate monitoring, the amount of data provided is around the 100,000 mark.

5 Heart rate variability (HRV) = sinus variability.

We also recall the principle of homeostasis: the sympathetic and parasympathetic systems are in a game of balance, the first acts as an accelerator and the second as a brake. This balance is an indicator of good health, which is statistically translated by a greater variability in heart rate.

6.4.2.5. *Statistics relating to physiological series*

The challenge, for the statistician, is the variability (of fluctuations) in heart rate time series. An illustration is provided in Figure 6.1 (120,000 data over 240 hours).

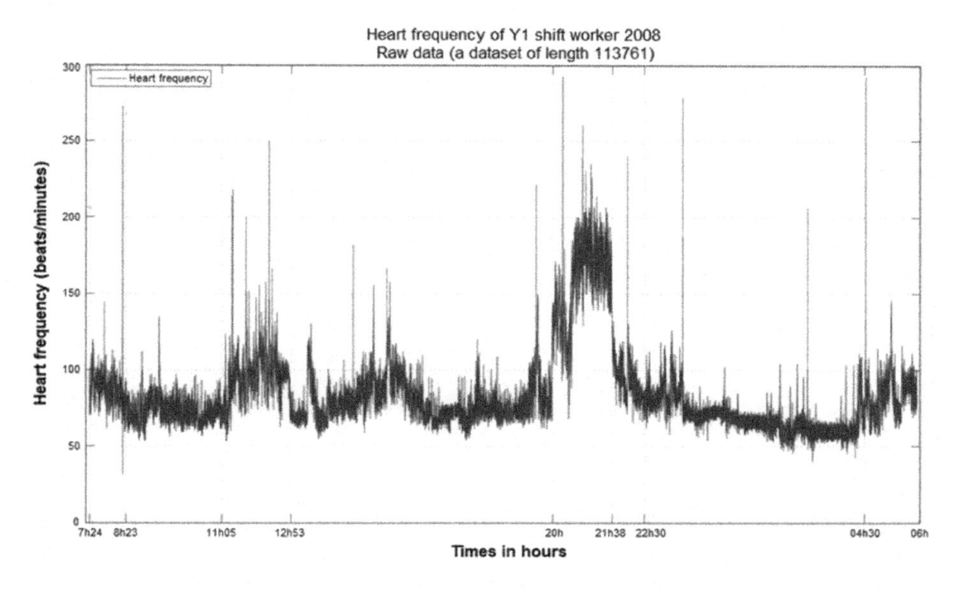

Figure 6.1. *Heart rate of a shift worker over 24 hours. For a color version of this figure, see www.iste.co.uk/picard/value.zip*

This is the heart rate of a shift worker (with assignments to different jobs) retrieved by the health department CHU Clermont-Ferrand. This shows approximately 120,000 heartbeats in less than 24 hours. The work shifts were from 6 am to 1 pm, with the afternoon free. The patient stated playing football from 8 pm until 9.38 pm precisely, going to bed at 10.30 pm and waking up at 4.30 am.

The sinus variability, in high and low frequencies, was also fluctuating. It is specified that this measurement is obtained in: (1) establishing the series of temporal differences between cardiac pulses, (2) making series of these measurements over time, and (3) establishing the corresponding frequency curve (spectral analysis using wavelets + change detection)[6]. The low frequencies are characteristic of the sympathetic system, where the high frequencies characterize the parasympathetic system.

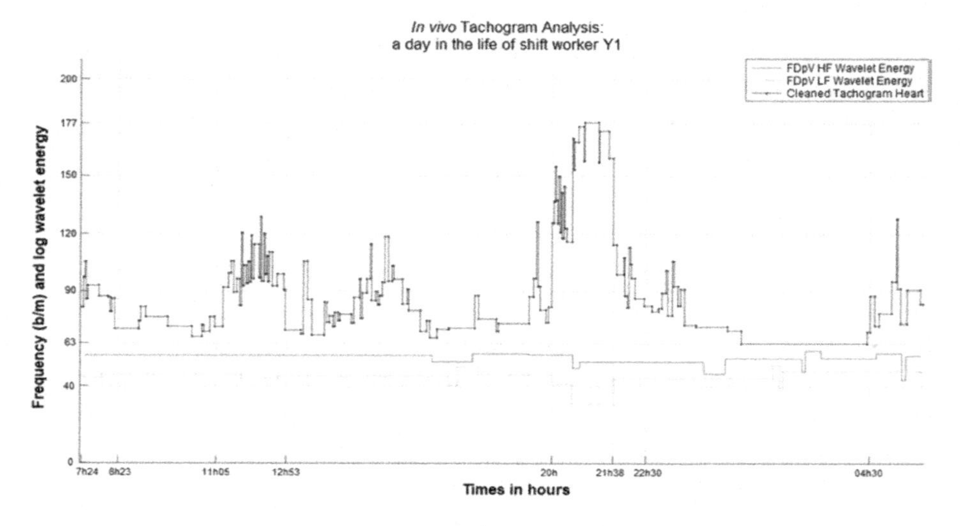

Figure 6.2. *Heart rate analysis in real-life conditions. For a color version of this figure, see www.iste.co.uk/picard/value.zip*

Schematically, the proposed system solves the limitations of current systems: (1) *in vivo* analysis of temporal series of heart rate and its regulation: 120,000 heartbeats analyzed in 15 seconds, compressed by a factor of 1000; (2) *in vivo* regulation (autonomous nervous system).

For the first case, we present a curve where we gradually detect a variation in heart rate during a more intense physical activity of the patient (football). The second objective is illustrated by a second curve in which the analysis of the regulatory system appears (sympathetic or parasympathetic). A decrease in the activity of the parasympathetic system is noted during the football game.

6 The use of wavelets is crucial for HF and LF analysis. Yves Meyer received the Abel Prize for wavelets. See https://lejournal.cnrs.fr/billets/le-mathematicien-yves-meyer-recoit-le-prix-abel.

This work was published in 2012 in an international journal [KHA 12].

6.4.2.6. *Summary: theoretical methods versus real-life situations*

Algorithms were developed to analyze physiological series with a computation time (and power consumption) being a function of n, where n denotes the amount of data analyzed. However, at this stage, there is not a reliable sensor providing continuous physiological signal measurements – those that are reliable enough to develop an application for the general public. A second limitation is medical validation. Such a validation (EBM or "evidence-based medicine") is precisely the purpose of the ANR "Do Well B." project presented in this section.

PART 3

Methods and Tools for Facilitating Appropriation

Introduction to Part 3

The real-life projects mentioned in the previous section indicate that a certain number of difficulties could lead to renouncing, at least partly, ambitions and valuable propositions initially promised by connected healthcare. These difficulties are of various natures: rejection by the user, patient or professional, high costs which do not allow the release of a viable economic model, regulatory obstacles, or simply the impossibility of making the solution technically work in a satisfactory way and at a reasonable cost.

To solve these difficulties, many tools, methods and skills are mobilized. These methods and tools belong to very different fields, including the human sciences, the law and the engineering sciences. Each of these areas carry their own vocabulary, culture and tools, and it is not always easy to articulate this expertise in a coherent and useful way, in projects dominated by the control of costs and delays.

This section aims to shed light on different useful areas, by showing how each of them is likely to create new solutions related to connected health. Without claiming to be exhaustive, here we will address issues of participatory design, legality, economics and some relating to engineering sciences. Specifically for the latter, a more detailed examination of approaches relating to connected objects has been carried out in the book *Connected Objects in Healthcare: Risks, Uses, Perspectives* [BEY 17] to which the reader can refer.

7

Design and Evaluation

Previous chapters have repeatedly addressed the issue of the place of the user in connected solutions.

It appears that this connected healthcare will have the greatest chance to work effectively if it makes sense for the human individuals who it links to, so that they encompass the challenges it carries. One way of doing this is to join up with representatives of various stakeholders to design and develop the components of solutions, through "co-design".

DEFINITION 7.1.– Co-design[1]: this process is defined as the execution of an innovation project in which a diversity of actors are mobilized to work on a common solution. The goal of co-design is to base the development of a device on a panel of fundamentally interdisciplinary actors. We therefore assume that the device will meet the different requirements having been co-designed according to habits, skills, visions and techniques specific to each individual. Co-design is concretely translated using specific methods (co-design workshops, ideation phase, etc.)

This co-design approach can be mobilized by a company that masters the corresponding skills and is able to mobilize the concerned ecosystem actors. But an alternative consists of calling upon specialized actors: the Living Labs.

Chapter written by Gaël GUILLOUX and Robert PICARD, with contributions by Sylvie ARNAVIEHLE and Perrine COURTOIS.

1 See the Glossary in [PIC 17a, p. 171].

7.1. Co-design and Living Labs

In a Living Lab, projects are always the fruit of co-design or co-development, if only because of the specific place occupied by the user (see the definition of Living Labs offered in this book's introduction, referencing the work of the Forum LLSA[2]).

The two books: *Co-design in Living Labs for Healthcare and Independent Living* 1 and 2 offer an in-depth presentation of the various methodologies which can be mobilized (book 1) [PIC 17a] from experienced examples (presented in book 2) [PIC 17b]. Human sciences in the areas of socio-ethnography, ergonomics and design are mobilized in this way. We give an overview of the approaches to ergonomics and socio-ethnography, illustrating them with short examples. The design approach will be further developed in the following section.

7.1.1. *Elements of method*

7.1.1.1. *First ergonomic approach*

In ergonomics, the importance of the "human factor" is illustrated on two levels:

– the acceptance of the device, when it is made available to the individual and broader users: this implies and exceeds the term "affordance". Affordance is effectively the ability of a system or product to suggest its own use. Affordance is linked to the understanding of the device's functioning, as the metaphor used during its design is meaningful to the patient. For the device to be acceptable, or even desired, it is also necessary for the individual to understand what the device brings to them, and for it to be attractive, for example, to have playful aspects;

– long-term adoption ("compliance"): this consists of preventing the user from abandoning the device ("drop out" phenomenon). For example, one of the main contributions of the sociological perspective implemented by the Diabète LAB is to explain the social, technical and cognitive mechanisms which cause patients to appropriate these devices or, conversely, abandon them.

2 www.forumllsa.org.

Concerning this second aspect, we have already mentioned the work of De Choudhury *et al.* on the predictability of whether a quantified self or m-health object will be abandoned [DEC 17].

7.1.1.2. *Socio-ethnographic approach*

The socio-ethnographic approach makes it possible to address the reality experienced by individuals in their environment, particularly that of chronic patients. The uses established around the appropriation of new solutions transform the various dimensions of the patient's work, the characteristics of their reflexivity and experience of illness, by introducing a new tempo in relation to themselves, their body and their pathology. This aspect is illustrated in the description of the Diabète LAB (section 1.3.1.2).

7.1.2. *Illustrations*

7.1.2.1. *One case: a connected pillbox*

In this case, the dimension of design and an assessment of uses is presented. The problem of articulating this type of evaluation with that of clinical trials will be discussed in detail later in this book.

Designing a connected pillbox

The connected pillbox is a connected object considered as a medical device (MD) in the regulatory sense. The fact that it is connected gives it a new and different value of use. Innovative features included in this connected pillbox are for example: the scheduling of when to take medication, with reminders (sounds/visuals on the pillbox, SMS on smartphones), the recording of the cell's opening times, recording medicine being taken, checking the regularity of dose administration, recording prescriptions, etc.

The project

The project for the development of a connected pillbox and its evaluation of use was financed by a French public fund, FUI, (*Fonds Uniques Interministériels* – single interdepartmental fund). Its initial product positioning was that of a pillbox to be used during clinical trials to control the actual administration of medicine. Ultimately, over the course of the project, this pillbox was oriented towards a mainstream use by the general public. It was therefore necessary to understand how such an object could be perceived, so that it becomes an object of choice for improving observance.

The pillbox included a box organized weekly (four doses a day over 7 days), a series of cells (blisters), a connected film glued to secure the presence of treatment and which could then detect a dose being taken. An application allowed us to set alerts, record delivered treatments and provide captured monitoring information.

It was a matter of integrating the patient into their ecosystem, consisting of carers (family), nurses, doctors and pharmacists who communicated through the pillbox.

It was also a question of ensuring the usability of the product on the patient side, who needed to be able to use it as a normal pillbox, with new functionalities such as SMS reminders, etc. In addition, all information was automatically recorded and was thus made accessible (if the patient had authorized access to their data by a relative or a healthcare professional); the protection of private health data was an essential parameter for this project.

Testing a prototype using the Living Lab

A major effort was needed to recruit professionals, particularly pharmacists. This was mainly because none of the evaluators were paid, regardless of their profiles. The project benefited from the support of the URPS (*Union Régionale des Professionnels de Santé* – French regional union of health professionals), pharmacists from Languedoc-Roussillon, as well as medical networks.

The recruited profiles included: pharmacists, nurses, doctors, carers and patient-users. In the Kyomed LLSA, workshops were organized for each profile based on test scenarios, using three or four workshops for each type of user (a workshop corresponding to a phase of the acquisition process/implementation of the object). Each evaluator was mobilized for an hour or two. To compensate for this short time, a film was presented to the evaluators to raise the question of uses over time. At the end of the course, a self-administered assessment questionnaire was offered to the participants. The most important questionnaire concerned pharmacists (54 questions).

Quantitatively, the test sample represented 96 individuals, more than half of whom were patients. The test population is representative of an object used by the general public: populations of all ages, all pathologies; pharmacists in pharmacies of varying sizes and doctors of various specialties.

Evaluation feedback

This feedback was comprised of several aspects: tips for users to be observant, according to profiles; identification of evaluators, potentially promoters, and how many there were (13% promoters in the sample); evaluators gave advice with the challenge being to bounce back and accumulate knowledge on uses for future products of this type. The importance of performing tests in early phases of development to form a new concept like that of the connected pillbox became apparent; it also consisted of anticipating the medico-economic analysis.

Regarding the product's evaluation, it focused on consumables, the blister, the film (used for sealing), subscription to the service (i.e. application use), data monitoring and so on. The challenge was to identify levers and usage obstacles: not only at the object's level, but also at the level of web application data monitoring.

It remains essential to monitor data in the long run. The issue of securing data was considered, but was not required for this project's evaluation, as personal data was neither recorded nor transmitted. Another lesson: pharmacists are ready to work with such a tool. But preparation comes at a cost, as it is very variable depending on the prescription, and it can possibly be very high.

7.1.2.2. *The Hadagio case*

The Hadagio solution has already been presented in Chapter 4 of this book. It concerns three categories of beneficiaries: health professionals (doctors, nurses, freelancers), patients (elderly people or those affected by a chronic illness) and social professionals (mutual insurances, associations). Here, we outline some aspects of the implemented design approach.

The user has access to services via a tablet and web access, and healthcare professionals and social workers can also have access via their mobile phones. An ergonomic approach has been mobilized with the participation of these various categories of actors: essentially doctors in hospitals at first, then many others, including patients and carers[3], and also adapted physical activity centers, doctors of various specialties: diabetes, cancer, gerontology, nurses and various freelance professions.

Focus groups have been put in place continuously, since the launch of the solution in 2012 (early development) and commercialization started in 2015. They include between 5 and 50 participants depending on the case and are made up of users from different categories, with the participation of experts in cognitive psychology and ergonomics. The attitude of the individual is analyzed, and their reflection is supported by visuals aids, like representations of scenarios for example. The movement of the user's eyes is monitored during the use of the interface, by looking at the individual's reaction to colors and pictograms. The ergonomic concern is allegedly major in this solution.

The solution is continuously evolving thanks to this co-design activity. This concerns an ergonomic plan, uses, and functional aspects. Older users are particularly required for a participative interpretation of the uses which make up the solution.

Box 7.1. *Co-design according to Hadagio*

7.1.3. *Patient involvement in co-design: areas of vigilance*

Often, the patient is solicited at the very end of a product's development cycle, to give their opinion, but no change can be made to the solution. This is what the expression "patient-alibi" refers to. It is important that industrial suppliers are aware that the value of the patient is potentially much higher earlier in the development stage, not only during the final one.

Other focuses of attention are to be taken into account:

7.1.3.1. *Selection*

We must pay attention to the way in which patients who will be partners are chosen: beyond the classical medical criteria required in clinical trials for the constitution of patient cohorts, behavioral aspects are to be taken into account such as empathy, listening, taking into the account the opinion of others, the ability to work in a group, to co-build, etc.

3 For example, two projects conducted with ADMR Haute Corse, funded by the Collectivité de Corse, to experiment with Hadagio, with the APA, mobilized speaking groups for carers and a system of serious games.

7.1.3.2. *Number*

The questions to be addressed are the following: how many users must be consulted? How do we ensure a good representation? How do we benefit from a wider vision? The definition of this number, again, is based on considerations that do not necessarily overlap with those concerning the selection of sample sizes for clinical trials.

7.1.3.3. *Training*

Patients who are trained to solve the problem are better able to play a useful role. Different areas are affected depending on the products and services, development processes, etc. The European Patients' Academy (EUPATI) provides patients with accessible and reliable information on how drugs are developed, but this only concerns the pharmaceutical industry. The patient who aims to make a valuable contribution to the design of solutions must know about toolboxes used by professionals, understand R&D challenges, development plans in biotech, etc. It consists of having an objective representation of things to be able to work with these companies.

Often, due to a lack of skills and preparation, the patient may limit their contribution to the signing of an attendance sheet, accepting everything offered to them afterwards (which refers to the previous notion of patient-alibi).

7.1.3.4. *Status*

Is the patient a consultant? Self-employed? Employed with a fixed-term contract, open-ended contract, associated with the operation of a project? The answers to these questions are also to be clarified.

7.2. Approaches integrating design

Today, design is subject to a certain level of infatuation. Public authorities recognize its importance, convinced that in global competition, design comprises:

> "An indispensable lever for the competitiveness of companies, whatever their size and sector (industry, crafts, services, digital, etc.)[4]".

The field of design not only concerns the aesthetic appearance of objects but also seeks an adequate solution to fit in with the future user. It is first and foremost a design process, centered on the user and the use of products and services. Living Labs are using it more and more frequently.

This is to ensure that objects and services will meet the requirements of future users, including emotionally, culturally or locally[5].

Several issues can indeed be avoided if the design of devices has been well led. This conviction is expressed in the case of Santé Landes (see section 4.3.2.1.2):

The devices could no doubt have been designed in a more relevant way and have been adapted the targeted situations: this reflection was not conducted within the framework of the program. The reason for this is that the TSN (*Territoire de Santé Numérique* – Digital health territory) program[6] is under harsh time constraints, and the mobilized objects are objects coming from the trade world. There has been no preliminary work on the emergence of new tools specifically adapted to the experienced situation.

A design approach ensures it considers the experiment in which the device must fit into, and which it must help amplify. The device's design must not be disconnected from it. This consists of "drawing" the whole experiment, specifying the human, geographical are technical contexts early on as well as the situations that will be encountered. It is important to put the stages of device design in a situation (with prototypes) to ensure their efficiency, and patient-user adherence.

7.2.1. *Design: a structured approach*

The steps of design-type approach, like the steps of co-design in general, are not standardized. The conduct of design projects proceeds with a

4 https://www.entreprises.gouv.fr/secteurs-professionnels/design.
5 [PIC 17a, p. 59].
6 The French "TSN" program is an initiative to test digital solutions for health.

dynamic that is unique to this approach, which we can prove by comparing a referential "design" approach (framed) with the operational approach that is presented in the following. It is important to note the significance given to experience, to empathy, and to understanding context, particularly environmental and social links. It consists of starting by understanding the user's experience to arrive at the target experience. Each operational approach fits, with its own codes and benchmarks, into the global approach carried out by Living Labs, more significantly those known as "ideation" and "prototype" [PIC 17a].

DEFINITION 7.2.– Design is a project activity which consists of determining the formal properties of industrially produced objects. By formal properties, we do not only mean external characteristics, but also the functional and structural relations which make the object a coherent unit from the producer's and the user's point of view. Even if the concern for an object's external quality allows it to be made more attractive and also conceals its constituent weaknesses, the formal properties of an object are always the result of the integration of various factors, of functional, cultural, technological and/or economical types.

A "reference" process from the academic world is defined as follows:

1) **Immersion**: observe the context and meet users to create a link between discourse and behavior (and better identify the implicit aspects that may impede a good capturing experience).

2) **Co-design**: co-create with patient-users and other potential stakeholders (carers, healthcare professionals, technicians, etc.). Mutualize skills which are technical, economic, healthcare related, etc., propose ideas and then develop concepts of future devices.

3) **Experimentation**: prototype and test concepts with patient-users and other potential stakeholders (carers, healthcare professionals, technicians, etc.). Validate a concept relevant in its uses.

4) **Development**: establish and implement the solution (the device).

Box 7.2. *Design approach: process*

To illustrate, in the following section, we explain the offer of a specialized consulting company, ELIA.

7.2.1.1. *Design: a market approach*

Preliminary step: engage with the patient

To begin with, this step will involve engaging with patients with early and effective communication. Before starting the project, the individual in touch with the patients will be able to ask the patient to participate in the process by orally explaining the benefits they will be able to obtain from the project. Hospital experiences are referenced, where this engagement takes the form of participation in workshops. We must ensure that the patients understand what is at stake. Different support measures can be developed to present the challenges and modalities of this participation and how to participate.

Step 1. Modeling user experience

This must be done in relation to the specific problem: for example, that of observance, the evolution of a chronic illness, etc. Field observations such as individual interviews with patients and their relatives are carried out. We meet the whole ecosystem. We obtain (1) a modeled, detailed and chronological description of the actions, interactions and emotions felt and (2) the questioning of patients.

An example of detailed experience mapping is shown. The journey starts very early, for example at the first sign of symptoms, or the day when the person took a risk. Then, all the difficult moments are reviewed. Curves and colors are used to highlight the points requiring attention and the most complicated moments: the use of a device, contact with a service (doctor, nurse, connected object), any area where things have not gone well. It is necessary to be very precise to bring out the real problem, the real risk. At the bottom of the diagram, the technical infrastructure elements mobilized through interactions are presented.

From this representation, an action plan will emerge. The proposed plan will be particularly rich. Simplified maps are also developed during the workshops with patients, which allow us to share a chronological vision with them.

Step 2. Identify areas of opportunity and challenge

It is with this narrative that critical points of the experiment are located and the related problems are identified ("spaces of opportunity"), to turn these problems into challenges to overcome. The method starts by breaking down the problem word for word from which we can draw "insight". The challenge is then drafted.

Step 3. Conceive of solutions/concepts

It is now necessary to conceive the responses to challenges without limits of feasibility and to summarize these answers in the context of daily use by developing scenarios. Design workshops are set up. There must be no limit, neither technical, nor medical, nor regulatory. We mobilize creative media and videos, for example, we can present a play. In the end, we have simplified prototypes which facilitate communication and decision-making, and pleasant stories which present solutions to the problems faced.

A particularity of the health sector is the following: in other sectors, such as distribution or banking for example, all actors are gathered in the same room. In healthcare, the doctor's authority makes things difficult. ELIA practice involves professional groups and patient groups separately. Working with laboratories, hospitals and MDs is more complicated, including at the decision-making level.

Step 4. Prioritize and plan a high benefit experiment

Project leaders meet in a workshop to define and rank the prioritization criteria (regulatory, technicality, feasibility over time, financing, etc.) of elements of the experiment. The elements must ensure that connected objects are well-used and adopted in the long term. We are not looking for a single solution, which would generally remain insufficient, but rather multiple smaller solutions and different objects. From these small solutions, we formulate one unique service offer. Then comes a phase of rapid testing involving users, it relies on various tools: comics, storyboards, stories and so on. Thanks to these preliminary tests, multiple iterations take place to improve the experiment, and a decision can be made regarding functional prototyping.

7.2.2. *Specific problem of the design of connected objects*

7.2.2.1. *Purpose of the object*

When designing a connected object, it is necessary to systematically research for whom the object is being invented, and work on its translation in the universe of that specific individual. There are two demands to satisfy: patient use, and also the insertion of the object in the prescription pathway. This is often what is missing and if this is not accomplished, it can cause problems.

In projects, when everyone expresses themselves regarding the objects they use, we often discover a multiplicity of actors and experiences. This must be undertaken upstream with extensive crossover work.

7.2.2.2. *Question of measurements carried out in real life*

The question of the object is enriched by the concept of connection and therefore measurements. The design approach for devices (products, spaces, connected objects) and collecting data poses first and foremost the question of what we want to measure. In design, measuring refers to facts and gestures, behaviors, uses and practices related to what we want to measure.

The key to success here is immersion with users and their observations, and the challenge is to determine the scenarios of uses associated with the device and the targets. This assumes:

– knowing what devices the users traditionally associate with (or will associate with) this type of measurement, or devices that they already use. If this is not the case, which devices will they associate these measurements with?;

– understanding the situations and the contexts that favor or facilitate measurement, notably in terms of space, schedule and pace;

– identifying the expected level of information and communication beforehand, during and after measurement. Emphasizing the notion of knowledge provided by simple information and access to data are also necessary;

– determining the minimum level of service expected to trigger an act of commitment and user involvement in the device.

7.2.2.3. *How will we measure?*

Scenarios of use identify the opportunities and limits used to define a device. Its daily use will be amplified by its efficiency and relevance, which can be introduced effectively and sustainably in these scenarios of use. These scenarios are built around answers to the questions: who, what, with what and how? They also need to define user–patient profiles and other stakeholders who interact with them throughout the scenario of use.

It is necessary to observe the services in a different way: what opportunities do they offer to user-patients and to other stakeholders? What are the limits?

It consists of integrating daily life to allow the formation of a routine, something easy and pleasant, a win-win situation. The device must participate so as to amplify and to have an experience (beautiful, interesting, motivating, etc.).

In addition, the materialization of the device is also one of the keys to successful experimentation.

7.2.2.4. *The device*

The device is understood in a broad sense: a system that is able to combine technical assistance, human assistance, services and activities.

To approach this, it must be designed in an articulation, with at least four components:

1) A spatial component: the space where the measurement(s) is (are) taken. In particular, it is necessary to manage the user's connection with this space.

2) A support component: digital media, products, furniture, equipment – elements that are necessary to capture data, the representation destined for the user, to professionals taking part at home and those conducting research (and other actors), and/or actions to be carried out as part of the capture on/with/by/for the support, to services destined for the user, to professionals taking part at home and those conducting research.

Ergonomics, forms, functions and aesthetics must amplify the experimentation support for users and all stakeholders. The design of tangible interfaces, in particular, is important.

3) A communication component: it is important to intelligently reflect on the identity, narration, formats and communication media such as that which is printed, filmed, digital, etc. before, during and after the experience of capture, representation and use.

The design interface and data visualization must be particularly well thought out to facilitate and accelerate the transfer of information, access to knowledge, avoiding interpretations (to be able to relativize the interpretation with access to raw data, for example), adaptation to cultures and to lexical fields of the recipients of the information, data and knowledge.

4) A service component: the services offered contribute to engaging the realization of the experiment before, during and after. It is also necessary to think about the articulation of the technical device along with the human intervention/help in defining experimentation.

7.2.2.5. *Articulating the device with its context*

The device is also thought of as a vector of links between the following elements:

– the collection area: behaviors, gestures and facts are collected and must be directly correlated with the measured consequence/effect. A counter-example exists, however: opening the refrigerator to know if the individual is eating properly. This act can be very different to the desired reality; the opening of the bin is probably a more relevant indicator of food consumption!;

– data analysis: the same data can be analyzed by several variants of algorithms, at least at the beginning of experimentation, and confronted with the realities of behavior and interpretation to avoid misinterpretation;

– the feedback of information or services to users who serve the experience and allow appreciation of adhesion and device acceptance.

Usage-oriented recommendations

The concerned device must, for the one who experiments with it:

– give the user the choice for action, to outsource or even allow them to adopt a laissez-faire attitude;

– be perceived by the users as co-constructed (the perceived engagement that the user has in a co-construction is not objective. Their real degree of implication to the co-construction is usually inferior to their perception of implication);

– make the user think they will gain a personal/individual advantage (evolution, progression, etc.);

– give the user the feeling that they possess some control regarding data flow;

– show that the technology that is imposed on them expresses with human interventions; we will avoid the feeling of dehumanized technological intrusion (see, for example, [TUR 11]) or that the object to be tested has some sort of consciousness, which can be frightening.

7.2.2.6. *Implementation*

The implementation can be ensured according to two different axes depending on the economic and technical capacity of the experimentation that must be conducted:

If there is an investment capacity in terms of time and cost:

It is advisable to redefine the device with the entirety of actors and in particular the manufacturer of the technology present on the market if required. This can simply require the integration of this technology in a new envelope – a disguise so to speak. 3D printing is relevant because it requires little intervention.

If the margin for maneuver is limited:

Emphasis must be placed on the experience provided by the service, the functions carried out and the human intervention, and not on the material support of this service. The latter must become anecdotal, as a simple support enabling people to have a desired experience. An existing device can be mobilized (if it is not possible to intervene in regard to the individual) all the while imagining a protocol/process of adhesion and experimentation that still allows a degree of autonomy (in the sense of liberty) of decision and evolution by the user. It is not the support that we experiment on but a function: the meaning shifts from the device as such to the service provided.

7.2.3. *Two illustrations*

The following two examples aim to delve deeper into the issues and methods used for taking into account the human dimension in design. The level of analysis presented is particularly acute and is reminiscent of certain socio-ethnographic approaches. The mesh of analysis really depends on the very nature and purpose of the products and services as well as the targeted audiences.

7.2.3.1. *First example: sharing personal data*

This is a two-step approach used to engage elderly individuals in their involvement and acceptance of a system for exchanging information from their daily activities (home data collection and feedback on information).

First step: empathy and observation of practices to design a coherent concept for existing gestures

As far as the consideration of context is concerned (empathy, observation and maintenance), this particular one involves meeting elderly people in situations of use with their entourage (family and friends) or medical professionals.

Existing connected objects are diverse. The first attempts of use showed that they did not fit into real life, the everyday and the object history of elderly individuals. Some of these elderly individuals even developed strategies to escape the surveillance that these objects provide. An example included the filmed use scenario created by Superflux where the elderly individual eats on one side, while manipulating a connected fork to indicate change.

An elderly individual is above all a person with their own habits.

Putting their behavior under surveillance tends to be a form of judgment, for example the individual is doing "badly" and they would need to "recover". This is especially felt since the purpose of the device is to warn the user of these "bad" behaviors. This kind of experiment does not go down well. A lot of individuals engaged in this type of experiment put an end to it quickly. If a sensor turns on, for example, the individual feels like they are being watched.

At this stage, the users were asked about their needs, and on the time they can dedicate to data capture. Other questions aimed to gauge the probable reactions of the other actors involved and their relationships with the people around them. There are nuances in the history of relationships that elderly individuals have with technology (e.g. a lot of elderly individuals can only be passive in front of a television – receiving information only). A pyramid of needs has been established to determine what will be maintained in the project. The assumption was that having a limited capture was better than having none at all.

This experience teaches us that there are two entry points to determine the reception of objects by the individual:

– the objects' security objectives: what is acceptable/accepted?;

– the emotions associated with their presence, which depends on their experience.

Most people were not ready to welcome a connection system with data to establish a linking system. It seems that there is a symbolic difference regarding relationships with technology. Raymond Loewy talks about the MAYA[7] stage, the acceptance threshold of innovation by a user, at a given time and in a given context. It is necessary to be able to create the steps that will lead the user to push back this threshold. It is thus a question of creating a connection and of finding the actions necessary to allow a first entry in communication via technologies.

The principle that led to our reflection at this stage was to design the device so that it induces simple and existing gestures. If there is a call, for example, we must be able to simply "take" that call or not – depending on if it is the doctor, family member, etc. – whether for video communication, listening, etc. The object must "create" a link with the application that is associated with it, with parents, friends or with caregivers. The relationship with the medical field must be simple.

Second step: to improve complexity of the concept and to change habits

A second, more evolved, object was proposed to users in a new step. This object had to be the carrier of further engagement. It was a question of finding a symbolic system, to give value to the relationship with the object, which had to "naturally" be an element of the interior of their habitat. It was imagined as a companion that was able to enhance the habitat, while

7 Most advanced yet acceptable.

requiring attention. It concerned provoking an engagement towards the object (whereas the object associated with the first step was only perceived as a communication tool). It associates a plant – which will have to be cared for, as it is beneficial to the plant – and a proposition of value for the individual, a personal benefit so to speak.

The concept integrates a new symbolic system of messages received by elderly individuals. For example, around the plant, the messages appear as different colored birds, depending on who sends them (family, friends, medical professionals).

7.2.3.2. *Second example: retaining a support link*

This second step concerns a research pathway conducted in the neonatal service at the University Hospital of Nantes in France. The team seeks to involve and recruit parents and their children to determine the consequences of diet on the appearance of a pathology (obesity, diabetes, cholesterol, among others).

This example is not from a Living Lab and differs according to certain aspects of experiments that are traditionally conducted there. On the contrary, it is particularly illustrative of the contribution of design in "scenography" and the "scripting" of a measurement experiment.

The neonatal service hosts parents and their children. One or two post-natal consultations are offered to them (between 0 and 7 years, following studies and research panels). The problem is that most families do not show up. When they accept to take part, the children can have tantrums, cry, etc. and often make measuring unfeasible. As a result, few protocols are completed.

Parents and children (three families) and medical professionals, are involved in the "immersion" phase. It turns out that the space is complicated: accessed only by elevator, because it is located on the seventh floor. Once at the seventh floor, the visitors have to call the protocol coordinator who leads them to a waiting room where they will later be picked up, etc. It is also necessary to work on the emotional dimension, allowing the children to have positive interactions, to give them elements of joy, of surprise, so as to obtain their engagement and make this experience motivating for them.

The scenario of use of the current situation is established on the basis of different steps and protocol sequences. It is initiated with an invitation letter to participate in the study, which is followed by a potential phone reminder. This is then followed by the family's arrival and the consultation of their child, until the resulting information is mailed to the parents, several weeks later. This all takes place over a three month period, but a card is addressed to the child the following year.

Many individuals do not show up. Many leave questions unanswered ("what do you do with this information?"). There is also a wish to integrate in the reflection the space in which the parents, the children and the medical professionals evolve, the equipment that participates in the experience of the course of treatment, and to better characterize the relationships between the various stakeholders. The fact is that the process puts the child (and the parents) under pressure and it is crucial to identify the critical points. On the basis of the critical points, seen as points of attachment for creativity, answers in terms of design attributes are identified, in connection with sectoral and non-sectoral monitoring: the level of personalization expected of the medical object; imitation, educational and simulation games; fictional material kits; form, ergonomics and aesthetics of certain objects, of spaces; flow and mobility, decor and ambiance, etc.

Graphs have been used to summarize data and to validate knowledge and findings of each step and with all of the stakeholders: timelines make it possible to describe the course of treatment in the terms of scenarios of use; stakeholder persona cards (expectations, behavior, obligations at each step, facts, observations, possible solutions). Following this first diagnostic phase, we move to a two-step stage: first, it consists of defining the child's positive and negative emotions (joy, playfulness, humor, sadness, etc.) and also identifying what leads to negative emotions. The challenge here is understanding how the experiment that will be offered is based on results based on good emotions. On the contrary, we aim to co-create in an imaginary way, with children and patients, the story of the journey, to capture different imaginations. Four stories emerge: the spaceship; when I'm older, I will be…; the star of research; and the pirate adventure. Work is then undertaken with a medical professional, and then with the mother and her child. This produces a validated scenario which happens to be a mix of spaceship and pirate adventure courses: the "explorer" scenario.

A survey is then engaged, to illustrate the story, by exploring the children's world and that of Jules Verne (since this is the theme that has already fed a first graphical reflection within hospital spaces). Graphical research is then conducted.

Finally, the scenario and narration that was proposed are accompanied by a story book. It accompanies the first letter received by the parents. It invites parents to read stories to their children. The story evokes in an imaginary way what the children can experience as part of research protocol. If they agree to participate (to experience the adventure), they receive a treasure map on which are positioned stickers representing the measurement that will be realized. The child can then reposition them in their story book, to continue their future explorer story. By removing the stickers, the child discovers a treasure map, their future journey within the neonatal service.

When the family arrives at the hospital, within the service, the child brings the treasure map. They find in rooms familiar elements to the treasure map that indicates the journey, and guides the child to where they must go. They become the guide within the service.

A requirement associated with these landmarks is that they are as cheap as possible: the solution was studied in a limited investment measure, in the realization and implementation of this decor: stickers and painted walls (with little decor but a lot of color). Each professional has got a badge representing one of the hero's stories, which the young explorer meets along the way.

A medal is acquired at the end of the journey.

The Bod Pod[8], a sort of box in which the child must enter, is one of the most critical elements of the journey (Figure 7.1). The Bod Pod is layered in stickers, making it less frightening, if not more familiar to exploration. The room is painted to simulate a seabed.

8 Measures body composition through plethysmography using air displacement in research, clinical or sport science.

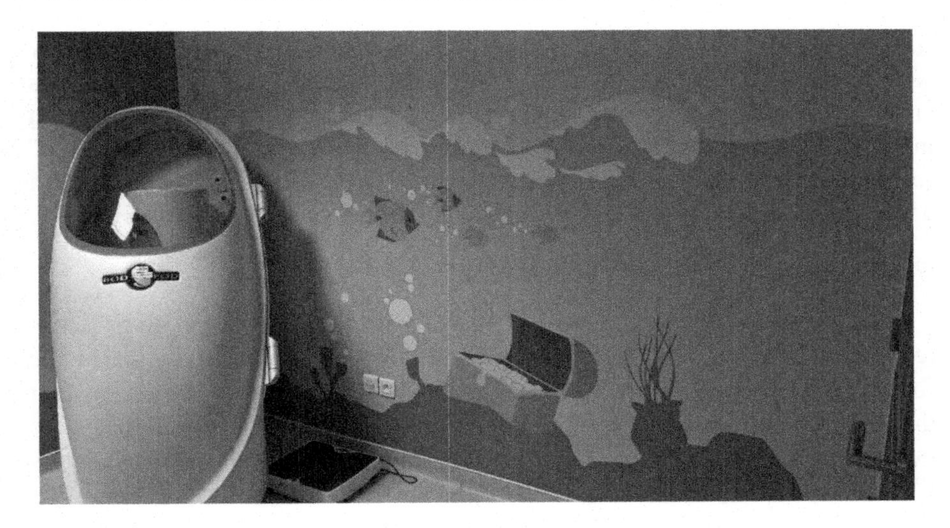

Figure 7.1. *The Bod Pod. For a color version of this figure, see www.iste.co.uk/picard/value.zip*

7.2.4. *The designer's role*

The three main roles of the designers are observer, facilitator and designer.

Role 1: observer (to identify the real question and the real problem)

The goal, for the observer, is to understand the user's experience and to reproduce encountered problems. The investigator proceeds via field observations and interviews based on a protocol. The experience can be reproduced in the form of a map of the experiences. Restitution is not a summary; it seeks a sort of completeness by pointing out the extreme limits of each of the behaviors. Empathy must be there, with a judgment based on their artistic and aesthetic position. This role can be filled in by and with other backgrounds, like for example sociologists, anthropologists, researchers, sectorial experts and prospectivists. Different perspectives can mutually enrich themselves.

Role 2: facilitator/mediator (of the disciplinary work)

The objective of the facilitator/mediator is to highlight the problem-solving opportunities carried/brought about by elements of the survey. They allow the co-construction of solutions with multidisciplinary work groups (ideation). The facilitator via the analysis of experiences animates ideation workshops, designs ideation exercises that they implement in successive sequences of work by mobilizing workshop materials: "pitch" formulation to relay experiences and their benefits in a few minutes, drawings, interviews, animation rhythms, methods and tools of creativity.

Role 3: designer (to bring about solutions)

The designers help the project participants to formalize their ideas (experts, users, etc.), then they transform these ideas into solutions in the form of experiments and test the experiments with the users to ensure that the functional/technical prototype properly meets specifications. The methods that they mobilize are paper, cardboard, foam, modeling clay, Lego, Playmobil, 2D or 3D models and in some cases functional prototypes. They provide the monitoring of the prototyping along with the technical teams for further development.

7.3. Conclusion

The elements previously mentioned show that the design is a structured activity in which the designer's role takes multiple forms. Under a unique term lies a complex integration of knowledge that is similar to the more codified and more academic knowledge of socio-ethnography, anthropology, psychology and ergonomics [PIC 17a]. The recent development of forms of design that explicitly integrated the social dimension (in terms of "social design") attests to this evolution.

The passion for design, evoked in the beginning of this chapter, must not hide this reality. Mobilizing an adapted design approach and a qualified designer to drive it is absolutely desirable, specifically in the health sector. But making relevant decisions remains tricky to this day.

8

Evaluations and Effectiveness

The question of evaluation is complex and multifaceted: assessing for whom, to what end and how? Medico-economic evaluation does not include user satisfaction; the evaluation of technical performance addresses other issues.

The case of mobile applications and connected objects was the subject of particular reflections and consultation at the public authority level, which justifies referring to it.

Living Labs include evaluation in their mission, particularly the evaluation of use. However, it must sometimes be articulated with a highly regulated clinical evaluation, which poses specific problems. We will discuss this point in section 8.2 of this chapter.

8.1. Evaluation of mobiles applications and connected objects

8.1.1. *Challenges*

We must progress in our knowledge of the impact of patient intervention on the R&D process in terms of time, cost and impact of the produced solution on patient health. It is indeed a complex question.

The question of "trust" is closely linked to that of evaluation. The recipients of solutions and, even more, prescribers and financers wielding public health responsibilities strive to ensure that solutions placed in the

Chapter written by Karima BOURQUARD, Daniel ISRAËL and Robert PICARD, with contributions by Hugues BROUARD, Virginie DELAY, Matthieu FAURE, Bastien FRAUDET and Thierry GATINEAU.

hands of citizens, patients and professionals are not carrying risks, and provide benefits to those who expect them.

To evaluate is to assign value. If this assignment is made by an actor who is not a stakeholder, who is independent of the offeror and the offeree and is recognized by both parties, the evaluation can constitute a solid basis of trust and consequently, market development, uses and clinical benefits.

Evaluating connected objects is important, because these objects are numerous and diverse, new, without previous socio-cultural reference, and leave the user, practitioner and consumer in a position of weakness. It is necessary to set up an evaluation framework which fits into a good synergy between the objects and the actions and behaviors of the individuals using them.

In healthcare, there is also the special consideration, expressed in the essential requirements of the European Community, relating both to the vital risks and to the fact that health status constitutes private information and privacy protection.

8.1.2. *Issues*

There are particular problems linked to connected objects (and mobile applications which are in a specific category, with a strong interactive dimension) and their evolution: their pace of evolution is no slower than other sectors – additionally, there is no determinism about the potential use of a given technology in healthcare. Their cost is particularly low as they fit in the dynamic of the general public and rely on widely distributed technologies. Owing to this, these objects are inexpensive and conventional methods of evaluation in healthcare, which involve lengthy and expensive trials, are particularly inappropriate.

These thoughts are independent of the regulatory time required to transform a quality standard into an enforceable requirement and this time is also totally incompatible with preceding characteristics. We will return to this topic in Chapter 11.

Nonetheless, many actors are expecting an assessment of these objects: those who finance and who search for valuable solutions; those who use; those who prescribe; and finally, those who manufacture and aspire to demonstrate the value of their product on an indisputable basis.

8.1.3. *Market responses*

This situation has led some private actors but also the public authorities to mobilize themselves to urgently answer this demand for reassurance. The question is difficult, and its analysis falls outside the scope of the current chapter. Regarding this, see Chapter 10 and [BEY 18] for a more in-depth discussion of the risk associated with connected systems. However, in this chapter, we mention two answers, one private and the other public, aiming to support market confidence.

8.1.3.1. *A private initiative (example)*

Harmonie Mutuelle, a French mutual insurance, offers complementary healthcare services, planning and savings and it has set up a "guide for assessing connected healthcare/well-being solutions"[1]. It is a major French player, supporting a population of 4.5 million French citizens (more than 10 million if we consider the VYV Group, of which it is a member), which intends to help raise awareness of issues and risks associated with connected objects and applications in healthcare among citizens. This mutualist actor thus manifests its concern to provide support for decision-making or for elements required to reassure the general public.

8.1.3.2. *Public authority's answer*

The public authorities addressed this issue more recently through a double initiative: that of the Comité Stratégique de Filière des Industries de Santé[2] and that of the Haute Autorité de Santé[3] (HAS) which we will develop hereafter. These initiatives have converged around a category of objects carrying health risks and strong developments (mobile applications and connected objects). The very short life cycle of these products makes conventional regulation models (regulatory texts) inadequate.

8.1.4. *The HAS approach*

The Haute Autorité de Santé wished to keep up with the rapid diffusion of mobile applications and connected objects with recommendations favoring the development of solutions which the users can trust. What is

1 www.guide-sante-connectee.fr.
2 Industry strategic committees have been set up by the French Ministry of Industry with the essential objective of defining and implementing a strategic roadmap shared by all players.
3 French National Authority for Health.

approached by HAS in its "References and guides for good practices in mobile applications and connected objects in healthcare"[4] concerns: the improvement of the quality of care, the quality of medical information, the patient's safety, evaluation in healthcare, coordination of healthcare and resulting interoperability, and finally, the question of cost-effectiveness.

What is being covered by other actors concerns the protection of privacy (by the CNIL) and questions of cybersecurity (by the ANSSI).

Questions of telecommunications, hosting and processing of data and reliability of algorithms are not discussed in this reference document.

8.1.5. *Evaluation matrix*

The variety of applications and their different levels of risk raise the question of how to weigh the criteria according to the main user and the declared main purpose of use of the application or connected object.

An evaluation matrix was made, crossing, on the one hand, the type of users (main target user of the evaluated application or object) and on the other hand, the "purposes of use". These established areas have an attributed level of criticality (color coded in the matrix presented in Figure 8.1); the most critical should receive maximum attention.

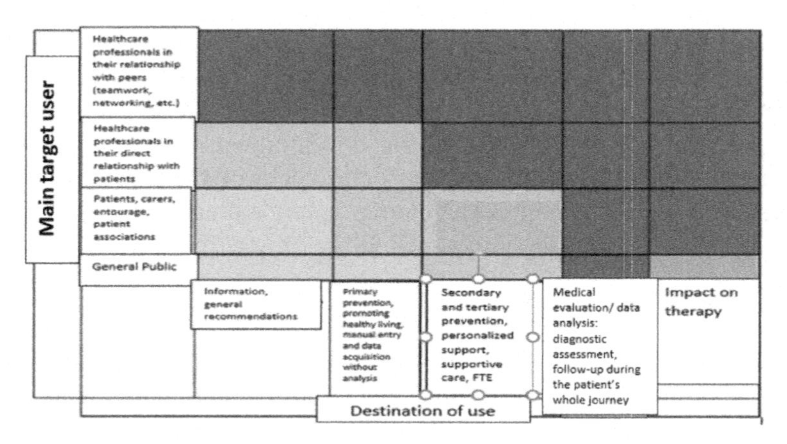

Figure 8.1. *HAS assessment matrix for mobile applications and connected objects*

4 https://www.has-sante.fr/portail/jcms/c_2681915/fr/referentiel-de-bonnes-pratiques-sur-les-applications-et-les-objets-connectes-en-sante-mobile-health-ou-mhealth.

Evaluation areas used in the HAS guidelines for best practice are the following:

– user information (description, consent);

– healthcare content (initial content design, standardization, generated content, interpreted content);

– technical content (technical design, data flow);

– security/reliability (cybersecurity, reliability, confidentiality);

– use/usage (design, acceptability, integration/import).

Concerning the evaluation criteria, 99 criteria were retained.

For example, the area of "user information" contains the following criteria: obligation of information; obligation of obtaining consent; and consent revocable at any time.

Each criterion is accompanied by a justification and an example. Each criterion is "desired" or "recommended" depending on the required level.

Box 8.1. *Evaluation areas of mobile applications and connected objects*

8.1.6. *Evaluation methodology*

First, this consists of finding the main target user (normally defined by the designer or manufacturer), as well as the application's or connected object's primary use (normally defined by the designer or manufacturer); the position of the application on the risk matrix makes it possible to select the requirement level according to the two previous answers (e.g. general information site for the patient corresponding to the green level).

At this stage, it is necessary to select the criteria for each domain and sub-domain according to the expected level of requirements (use of the spreadsheet provided with this document or the summary table of each domain in the previous chapter); criteria that are not adapted because they do not correspond to the specificities of the application or connected objects are excluded (e.g. the criteria corresponding to the reliability of measurement collection are not usable to evaluate an application of information in healthcare); the evaluation of criteria considered to be "recommended" and "desired" is then conducted. Certain criteria require establishing a risk analysis which cannot be conducted without support from competent individuals; finally, the results of the criteria evaluation are compiled and a summary is produced.

8.1.7. *Perspectives*

One difficulty concerns the evolving nature of this market.

Another issue is the question of how the applications and objects will integrate into a larger architecture. It is perhaps more important to evaluate the entire system instead of a single connected object, which is just a component. One answer is that the HAS guide does not deal with the technical "telecommunications" axis, and therefore cannot tackle the entire system.

A criticism can be formulated – without questioning the relevance of the grid – on the complexity of this result: in practice, it will be difficult to carry out all the required analyses in fields as varied as the object, application, data management, interactions, etc. Moreover, if this control is exercised on all these dimensions, there is a risk of impeding innovation. Too often, too many guarantees are required. It would be better to leave a margin for designers. In response to this argument, it has to be considered that the document produced does not act as a regulation but a guide to good practices. The idea is to support innovation by providing some basic elements but they have no mandatory status.

8.2. Evaluation of use versus clinical use

Designers of technical objects, connected or not, and those who design digital tools such as mobile applications, add an element of security to secure their projects by using a process exploring the needs of design upstream in the cycle. However, this exploration work is not to be confused with the work of UX designers who design the experience of using a tool (interface) with developers.

When products possess the status of a Medical Device, an additional element of complexity is added, because of regulatory procedures. This also has an impact on the economic model, a dimension which is also normally addressed in the process.

The case of the MoovCare solution (a web application used for personalized clinical monitoring and early detection of relapses and complications of lung cancer presented in section 6.4.1) illustrates the nature and challenges associated with these clinical trials.

8.2.1. *Clinical trials and their challenges: MoovCare, the connected solution*

Three studies regarding this solution were conducted successfully.

A first study, based on 43 patients over a year (2012), made it possible to construct and validate the specificity and sensitivity of an algorithm regarding 6, then 11 clinical symptoms, with identification of relapses 5 to 6 weeks before the scanner. The results were published in 2012.

The second study, also mono-centric and prospective, was launched in 2013. A total of 42 patients were included, afflicted by primary lung cancer (stages III and IV) and had to be followed with the solution. The evolution of these patients was compared to that of 49 patients followed in the traditional way. The observed gain in survival rates was very significant: 27% improvement in the survival rate between the two populations.

The third study aimed for a high level of proof. It consisted of a randomized multicenter phase III controlled study. Recruitment involved 133 patients, of whom 121 were eligible. They were divided into two equivalent groups: one monitored by MoovCare and the other by classical methods (visits and scans). The main evaluation criterion was the measure of survival, and the secondary criteria concerned the identification of relapses by the practitioner, quality of life, PFS[5] and reduction in the cost of care.

The comparison of costs deserves attention: while the ICER – Incremental Cost Effectiveness Ratio – represents several hundred dollars, that of the solution is less than $10,000, and the benefit in terms of survival is 30 weeks (1–4 in other cases).

The third study confirms the improvement in survival rates, of the order of 26% at 12 months.

5 Progression Free Survival: generally defines the amount of time between the start of a clinical trial and either the onset of disease progression (evolution of the pathology) or the date of the patient's death.

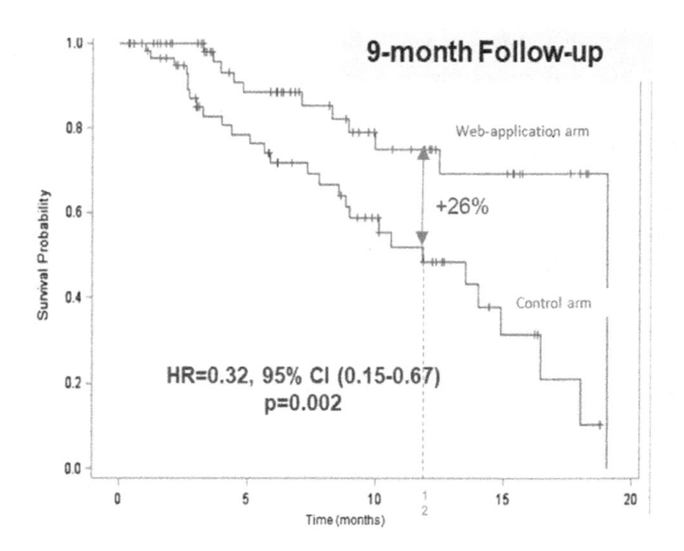

Figure 8.2. *Comparative evolution of survival rates with and without MoovCare*

Dr. Fabrice Denis, MD, PhD radiation oncologist in France, presented the results on Friday 1, June 2018 in an oral session in ASCO, Chicago. MoovCare™, a web-based patient-reported follow-up solution, demonstrated remarkable results in the 2-year follow-up: 50% of the patients were still alive when followed by MoovCare™ versus 34% with standard follow-up. These results confirmed the preliminary results that were presented at ASCO 2016 in a podium oral presentation and published in the JNCI 2017.

The survival advantage is explained by MoovCare's early detection of relapse at a time when more than three quarters of patients have good physical function, compared to standard approaches to relapse detection when only a third have good function ($p<0.001$). Earlier relapse detection when patients are healthier enables initiation of an optimal cancer treatment – which was feasible in 72% of MoovCare patients but only in 33% of standard care patients ($p < 0.001$).

The design of solutions for the patient at home also requires carers and professionals who take care of them. The question of acceptability is multifaceted. The design of solutions associated with the territorial platform of Santé Landes (section 4.3.2.1) is again another context, since the actors are mainly from the professional field and the support staff associated with the platform. The patient is not directly involved.

8.2.2. *Evaluation of the LLSA's approach*

The LLSA's global approach is simplified into four main stages:

1) framing, population approach;

2) ideation;

3) *in vitro* evaluation;

4) evaluation in "real life".

Each of these major steps carries an evaluative dimension: the first allows the formulation of ambitions; the second step is the site of formative and recurrent evaluation; the third pursues this work but concludes with a summative evaluation of the solution; finally, the fourth renews this evaluation work but enriches it when the time comes to measuring the impact. In terms of healthcare solutions in the medical field, and more generally devices labeled as medical devices, this impact includes a medico-economic dimension.

8.2.3. *Conjugation of LLSA evaluation and evaluation associated with clinical trials*

Evaluation in a Living Lab gives a major place to the individual and collective use value. Medical proof (medical benefit, medico-economic value) is not necessarily required. Conversely, in clinical trials, explicit use-based requirements focus on the ergonomic dimension, which is limited to direct users.

However, the sensitivity of individuals and professionals to the dimensions of individual and collective appropriation of tools is growing. The clinical outcome of a solution depends at least as much on this factor as on technical or medical parameters. The desired value of a health solution can no longer stop statistical proof resulting from an intervention, but will be enriched with elements from real life and also over time.

Thus, whether it is a connected object intended for the "well-being" market and also claiming to possess a medical value (e.g. post-clinical validation), a medical device in need of performance upgrades to convince the recipient-users (corrective co-design), or of solutions seeking to maximize these different dimensions (co-design and joint medical evaluation), a multidisciplinary method of evaluation seems necessary.

Three typical cases are briefly presented below, without pretending to cover the exhaustiveness of situations.

8.2.3.1. *Corrective co-design*

An MD requiring an improvement in its value of use can be undertaken using a Living Lab approach, which will integrate, in terms of population needs, the identification of the ecosystem concerns, the users of the solution as well as the criticisms and difficulties that the users formulate.

Inclusion criteria will be taken into account, as well as more broadly the configuration of clinical studies that led to marketing authorization. The hypothesis of a phase IV or post-marketing test should also be considered early on, to avoid any need to adjust the results of the LL approach during these tests, a source of delay and additional cost (especially in terms of data collection).

8.2.3.2. *Ex-post clinical valuation*

A technological solution for health (living well, helping with daily life, monitoring physical activity) can have some medical potential and lead to the manufacturer or publisher claiming MD status. Living Labs have allowed the realization of technical and usage files, in terms of prototypes or real-life products, potentially a collection of data of diverse natures. It is appropriate that these data, configurations and results of clinical tests are available for the clinical study's principal investigator, who takes them into account when designing their study and identifying the elements necessary for the completion of the pre-clinical and then clinical study.

8.2.3.3. *Co-design and joint medical evaluation*

In this case, co-design as well as formative and summative evaluations of the solution are seen from the outset as contributors to the intervention and are placed under the responsibility of the principal investigator. At least one secondary objective concerning the issues of acceptability and/or organization will be associated with this investigation. The prototype of the solution must meet the requirements controversially associated with a demonstrator from pre-clinical trials. The populations (patients, professionals) associated with co-design will be identified and included with those of clinical trials, without excluding the participation of other categories of people concerned even indirectly by the solution and possibly not associated with the tests themselves. These choices are negotiated with the principal investigator who ensures consistency between the choices and intervention, of which the architecture can be influenced.

Two cases are presented below to illustrate this point.

8.2.3.4. Two *case studies*

The following case differs from the one presented in Chapter 7 (see connected pill box, section 7.1.2), broadening its functional scope: not only is it not the same object, but also the intended evaluation clearly includes the regulatory concern.

8.2.3.4.1. Case 1: thess – the connected MD solution of Stiplastics Healthcaring

In this project, the targeted solution is one of securing and dispensing oral medication, monitoring administration and clinical data, in real time, for cancer patients treated at home. It is a connected medical device, with a digital interface for monitoring and with security constraints in real time. The determination of the class of this medical device will distinguish the requirements associated with that of the solution.

Among the objectives are: reducing iatrogenic risks thanks to prediction tools; improving coordination by means of the patient's relationships – doctor, city, hospital – and involving the patient through easy-to-use web services.

It is also necessary to reduce the overall costs, acquiring better control of environmental costs and improving the predictive capacity of the system by regularly monitoring in order to reduce the dosages. It is better to adapt the continuous dosages than to initiate, if necessary, sudden stopping of treatment.

Depending on whether there are one or more drugs involved, a simple dispenser will be available or, on the contrary, it will be necessary to introduce intelligence, with the possibility of a "go/no go" prescriber and notification of drug administration. A cloud infrastructure will be mobilized for data exchange. The development includes decision support software and algorithms.

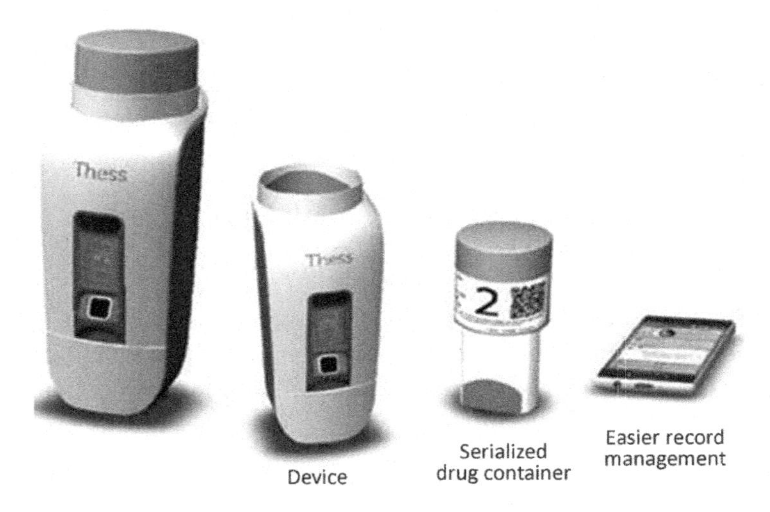

Figure 8.3. *The Thess solution*

This is a complex system with many interfaces, which poses from the outset the problem of acceptability and usability.

A prototype will undergo experimentation in a hospital setting, in two distinct centers (2018). It will be a question of checking the acceptability of the device, software and services. There is no absolute truth in this matter. The oral decision requires a multidisciplinary consultation meeting (*réunion de concertation pluridisciplinaire*, RCP), which specifies the protocol to be followed. This determines the follow-up. In oncology, in many cases, therapy must be adjusted to reality. Moreover, the ergonomic issues of ease of use must be considered for each actor. A financial return will be realized: hospitalizations, stopping treatments, modification of treatments, etc.

The organizational dimension is also discussed, particularly with pharmacists. This dimension concerns the logistics of the delivery of devices, the preparation of blisters for each week, and the management of empty medicine boxes that are the property of patients usually stocked in pharmacies. This was handled by workshops and a questionnaire.

The second case presented in the box below illustrates the necessity of interactions between the user and the technical device during its design, prior to its clinical evaluation when this interaction is, in a way, the solution's "active ingredient". The aim of this device is to support the activity of patient self-rehabilitation.

8.2.3.4.2. Case 2: BYMTOX – a serious game of rehabilitation

To this day, patients invited to practice self-rehabilitation are given a paper booklet. We observe that their motivation diminishes after a few weeks.

The support tools for the continuation of exercises are therefore not very motivating and do not support observance in the medium and long term. This observation agrees with the data in the literature [BON 13].

The project presented in this case, named BYMTOX, aims to evaluate the effectiveness of the botulinum toxin – a self-rehabilitation program using innovative support in a stroke population.

BYM presented a portable application architecture consisting of an iPad and sensors. The first version of the application targeted a single exercise. The content was enriched incrementally with upper limb rehabilitation experts at the Saint Hélier center. The solution was then tested with target patients of the solution: hemiplegics treated by the center. This allowed for a first level of feedback, giving an idea of the tool's potential.

Some of the patients involved in the test were enthusiastic. One of them said: "Figure it out, I want one when I leave".

The content was corrected, in terms of form and substance: speed of scrolling, sequences and duration of exercises, management of breaks, consideration of hemispatial neglect, fixing of sensors, etc.

After 6 months, an operational prototype of the solution was available, the "BYM Kit", the first serious rehabilitation game for mobile devices. This prototype had the following characteristics: user feedback, a playful and motivating interface, adaption to the targeted audience, information provided to the user about their performance.

The next step involved a pilot study, a comparative study to evaluate the effectiveness and usage of the tool on a "difficult" population in an ecological situation (outdoors). It concerned a stroke population in a context where they had just received an injection of botulinum toxin (limitation of muscle stiffness) and were not interacting with the hospital. This choice resulted in high stakes for self-rehabilitation, and from the desire to be able to measure the contribution of the tool to motivation – and so to observance – of this population.

The objective of the clinical study was to compare two populations belonging to the mentioned target (stroke + injection b.t.): one equipped with the tool (iPad + sensors) and the other equipped with the classic paper booklet.

The interest of the innovating tool compared to the booklet concerns the possibility of feedback to the user in relation to their action, the fun approach ("serious game" interface, evaluation of positive performance) and the generation of statistics. However, the programs are identical in terms of content: same exercises, same sequences and same daily amount. The study was conducted over 3 months, with 20 patients for each of the two configurations (booklet vs. tool).

The evaluated parameters concern the function of the upper limb, its spasticity (stiffness, contractions) and the motor skills of the limb. However, they also concern the effect of the tool: observance, motivation, satisfaction and interest of users in the usage of the tool.

8.3. Test beds: presentation of the work of the Forum LLSA's "test bed" group

A work group[6] has been formed within the Forum LLSA to share members' experience in terms of managing this type of platform. The origin of this initiative stems from the necessity to respond to a demand by the European Institute of Health (EIT Health) to document the "Living Labs and test beds". However, in EIT Health's documents, the test beds are only briefly described as "Living Labs without the participatory approach of users".

---

6 This group is different to that working on "connected health" presented in the introduction and of which this book is the result.

The work group was therefore formed to provide both feedback, as with all the groups of the forum, and a precise definition of this concept. K. Bourquard led this group.

8.3.1. *Concepts and definition*

Both test beds and Living Labs refer to the four-step co-construction process codified by the Forum LLSA: analysis of needs, co-design, prototypes and evaluation in real-life circumstances. Test beds are specifically mobilized in the last two steps.

The first definition used by the group for the test bed is as follows.

DEFINITION 8.1.– A test bed is a test platform in a controlled environment that allows the realization of replicable, automatable, reliable and rigorous tests to verify the conformity of prototypes, products or services with previously defined requirements.

Literature research supported the work of finding a definition by providing additional insights. This research and the exchanges within the work group make it possible to clarify the following statements.

"The test bed is an area equipped with instruments, etc., used for testing machinery, engines, etc., under working conditions"[7].

The test bed is not a substitute for a Living Lab, but it is a major instrument whether or not it is integrated.

The test bed is a controlled environment as close to reality as possible allowing objective measurements to be undertaken, to be collected, analyzed and reported following a pre-established protocol and allowing the reproducibility of tests.

Its purpose is to verify whether the formulated requirements are properly implemented in the object of study and the requirements can vary depending on the tested solution. Similarly, the test methods depend on the object of study; the group agreed on the following definition of the test process implemented in a test bed.

---

7 Collins English Dictionary.

– A formalized test process described by the test strategy, test methodologies (sampling, choice of test tools), test plans, test games or sets defined based on the requirements or specification of the object studied;

– test operators or a test team with the appropriate expertise;

– automated or non-automated components called test tools (sensors, monitoring, simulators or validators, processing or calculation tools) allowing objective measurements; questionnaires (audit) can also be used allowing the statistical analysis for the analysis of qualitative behavior;

– traces or logs and their conservation;

– validation reports provide the test results for the studied object;

– end-users' in-use situation of the studied object.

Box 8.2. *Characteristics of the test process in a test bed*

The organization of the tests is based on quality procedures, composed of structured approaches:

– standardized processes applied to tests;

– standard-based frames of reference;

– test platforms and test tools;

– objective validation criteria.

It takes into account the contexts of the users and the regulations in force.

During the group's work, the test beds in use within the Forum LLSA, and also eventually outside, were searched for and contacted. An initial list of seven test beds that responded and provided the necessary information was established. It was intended to extend to all the test activities present in the forum.

Currently documented are: Ngagement by b<>com (Rennes), Connected Health Lab by Isis (Castres), Gazelle Test Bed by IHE Europe (Rennes), Prometee by Telecom Nancy, STREETLAB by Institut de la vision, Paris, the CIC-IT Evalab by CHU Lille and the Guide de la santé connectée by Harmonie Mutuelle, Paris.

For each of these test beds, applicable standards have been identified. A total of 21 standard frames of reference have thus been identified, in very diverse and complementary fields. All of them are potentially requirements for evaluations.

A test bed can be hosted in a Living Lab and it can work for one or more Living Labs. There is then a potential for mutualization in this area.

The offer of service of the registered test beds is the following (end of 2017):

– b<>com *Ngagement* within the Usage & Acceptability lab (UA) of b<>com conducts a continuous development of new tools and methods to measure user experience, based on instrumented recordings, artificial intelligence, the analysis of immersive interactions or even the longitudinal evaluation of acceptability;

– Connected Health Lab: the strategic position of CHL is the innovation in connected health regarding patient pathways (pathway of coordinated care, healthcare pathway, frailty pathway, etc.);

– Gazelle Test Bed: interoperability and compliance of information systems, mobile applications or devices in the field of e-Health and AHA (Handicap and Autonomy) with respect to profiles and standards;

– Prometee: measure of the subjective quality according to uses in medical imaging;

– STREETLAB: evaluation of products and services for improving the visually impaired's level of autonomy. Evaluation of the therapeutic benefit of medicine, re-education and rehabilitation program;

– CIC-IT/EVALAB: validation of the usability of innovative new technologies for health and medical devices (MDs) – or new versions of such technologies or MDs. The validation of usability results from the summative evaluation of usability. It only concerns the final versions of technological solutions for health and MDs, namely versions ready to be placed on the market (excluding prototypes and other trial versions);

– Guide de la santé connectée: multicriteria evaluation of connected objects in health (reliability, manufacturer commitments, user reviews). The evaluations focus on health benefits and vigilance and not on the technical features. Two types of connected objects are evaluated: well-being objects (trackers, scales, sleep trackers) and MDs connected to self-test use (sphygmomanometer, glucometer, etc.).

8.3.2. *Perspectives*

The test beds are currently poorly known, and it is necessary to make them visible. It should be noted that there can be several test beds in a Living Lab, but also possibly none: in this case, the Living Lab relies, where necessary, on external test beds, independent of the Living Lab.

The identification and characterization of test beds also pointed out challenge professionalization. It is also a question of defining their field and their mode of intervention. Progress in this field can benefit both test beds and Living Labs.

The implementation costs of test beds are not published. This is due to the activity's highly competitive nature: mastering test protocols and publishing them involves controlling market access. However, we suspect that there is an interest for a Living Lab or an industrialist to solicit available test beds if they already exist rather than building a new one. The community has an interest in taking advantage of what is available by contributing to the economic stability of these tools. This is the interest of a test bed repository. Indeed, referring to a standard in test protocols can be relatively costly for a test bed.

There are certainly test devices that could be part of this approach, provided that protocols used are formalized and published, and ideally that they are promoted as international standards. It is necessary to develop exchanges between ecosystem actors to allow new collaborations in these subjects. In this respect, the usefulness of a test bed repository is also the localization of skills: it is necessary to have the email addresses of experts responsible for these tools in order to be able to exchange and benefit from their advice on the procedures, experience plans, etc.

As far as the testing of autonomous connected objects is concerned, experience shows that they are specific to each object and require a specific test protocol. For the case of the sphygmomanometer, for example, the connected sphygmomanometer is placed on one arm, and a classic sphygmomanometer on the other. Several measurements are made by exchanging the arms, and then the averages can be calculated. However, other protocols need to be invented for other connected objects.

9

Economic and Legal Aspects

In many phases, during the life cycle of connected objects, the economic aspects are tied to legal aspects and vice versa. Therefore, both these aspects are grouped in this chapter.

9.1. Regulation in real life

Regulatory obligations have repeatedly been referenced as barriers to innovation processes, sometimes even when they are not in the type of situation that leads to regulation. These difficulties are illustrated by the following examples:

9.1.1. *The connected pillbox*

This type of solution clashes with regulatory limitations:

– as soon as the project was launched, there was a prospect of authorization to unpack the drugs, so only the needed quantity was delivered according to the prescription doses to administer. Packing was also regulated; however, this did not occur. Thus, it would be useful to change the rules that concern conventional packaging;

– however, sale by the unit is also not authorized, except in rare situations. This reduces the possibility of pillbox use, particularly preventing the pharmacist from preparing blisters for a week's treatment. It curbs the emergence of personalized and intelligent packaging;

Chapter written by Anne-Marie BENOIT, Myriam LE GOFF-PRONOST and Robert PICARD.

– if the patient's commitment is the key to good observance, the pharmacist is the guarantor of the drug's delivery; however, in practice, the nurse often intervenes, and their work is not codified.

All this limits the pillbox's possibility of use, which can contain several medications. Moreover:

– the value of the doctor's follow-up evokes the topic of side effects: beside measurements, we could have imagined that the exploitation of information circulating would occur. In fact, this is not possible. The reason being a regulatory one: what information we want to recover must be consented to by the patient. However, we do not control what is exchanged on social networks. But anything that is measured or collected by the patient is exploitable. We define categories, we proceed to classify by cluster, which makes it possible to adapt care to each category. In collected information, there are visual analogue scales, which are used to guide the patient. Note that the data exploited by the doctor is hosted on a physical platform separate from the platform accessible by patients;

– in a context where the regulatory requirements are greater, where the application's software itself acquires the status of a medical device, a change in algorithm will require the product to be certified again. This will be an obstacle to innovation, costs and additional delays.

9.1.2. *The E4N cohort (section 3.2)*

The mobilization of connected solutions in this project also meets regulatory obstacles. Thus:

– in home monitoring systems, a strong authentication system is needed. An attempt was made to use the veins in the patient's hand to authenticate the patient. This hypothesis was categorically refused by the CNIL. Therefore, everyone can be authenticated, except the patient. The question is how to do this for the E4N cohort;

– indeed, this is a complex topic, and has caused project delays due to regulatory reasons. More specifically, it is at the level of the CCTIRS (*Comité consultatif sur le traitement de l'information en matière de recherche* – Advisory committee on the treatment of research information) that this delay is caused. The problematic part was that information was about families and there was genetic data and sensors that were used.

To overcome this, it was necessary to explain the specific scientific purpose of population cohorts of very large sizes and to manage them in the long term. This allows us to observe the general population and makes it possible to apprehend information which the academic and industrial world needs by looking ahead at the next 10 years. For this, what is happening today must first be understood.

9.2. The cost of regulation

Regulation sometimes has a cost, for various reasons illustrated in the presented projects.

In the MoovCare solution, there is a desire to keep all collected data, which assumes it will be anonymized. To do this, in France, the anonymization algorithm must be approved by the CNIL, the French Data Protection Authority. Certification takes 18 months. During this time, patient data are lost, which creates a major disadvantage: indeed, the idea is to keep records of each patient's health week after week until their death. These data do not exist today. We could measure the effect of other treatments, of a molecule, etc. on a certain type of patient. What will it be in 18 months' time? There may be other solutions on the market which are not subject to such harsh regulations.

Another handicap of the French context is that respecting regulations surrounding the hosting of health data multiplies the cost of this hosting by about five times.

9.3. Soft law: an answer?

Economic agents, manufacturers, law professionals and citizens share the observation that the "time for innovation" is not always "the time for law". In the eyes of most of our contemporaries, the norm of positive law would not allow us to better understand the economic, sociological and legal realities and also technologies, due to their rapid evolution which would profoundly transform our economy and social relations. In this digital society, constraints imposed by legal rule can indeed sometimes appear to be unadapted to the competitive and financial imperatives of companies, and to the needs of customers and users. How can legal security be ensured for companies but also for consumers in an increasingly competitive world?

Could soft law plans such as codes of good practice, professional recommendations or references provide part of the answer? When it comes to soft law, should this "movement" be considered a fad or should it be seen as a substantial one?

In addition to these imperative rules characterized as "hard laws", we can see the emergence and recognition of "indicative rules". Charters of good conduct, references and guides for good practices, and recommendations in rapid expansion, attest to this evolution. These deontological codes and these references most often designed by the professional fields thus designate a set of behaviors considered as indispensable for their practices by most professionals. Sectorial in principle, they can be content with recalling legal obligations. They also aim to overcome these obligations, to "adapt" them to professional settings[1]. They also participate in the regulation of a sector's activity. Often established in the framework of a qualitative approach, they are nevertheless only opposable to third parties if they have been published. Thus, these writings are also indicators of what should be done.

Through these codes and recommendations, public and private actors become new sources of law. These devices are successfully multiplying and producing norms, even though they detach themselves from the concept of sanctions, in all branches of law and particularly those related to innovative technologies. The "connected healthcare objects" whose purpose is to collect and process personal data are particularly well-suited to these new practices. Soft law then becomes the very expression of regulation.

The term "soft law" has its origin in 1930s international law where this "soft law" was required, in opposition to hard law as a means of regulating relations between states. Nowadays, soft law concerns many sectors in the financial and economic sphere (banks, agribusinesses, IT, etc.). In direct relation with the development and respect of soft law, an entire ecosystem is being developed. New professions appear, for example the compliance officer whose mission is to ensure company conformity, its managers and employees' behavior to whom legal and ethical norms apply.

1 See guidelines on connected healthcare: https://www.has-sante.fr/portail/jcms/c_2681915/fr/referentiel-de-bonnes-pratiques-sur-les-applications-et-les-objets-connectes-en-sante-mobile-health-ou-mhealth.

The efficiency of soft law is the result of a long process. Three reports by the Conseil d'Etat attest to the doctrine's evolution. Since 1991, the Conseil d'Etat has been lamenting in its annual report [SÉN 18] on "legal security", that the heterogeneity of the normative measure is the source of legal insecurity. The Conseil d'Etat states that "the law no longer appears as a form of protection but rather as a threat". The soft law approach then begins to be conceived as a means of supporting open innovation. The second report [CON 05], on legal security and the complexity of law, published in 2006 argued the fact that there are many measures which originate from law but are nonetheless devoid of legally binding forces (codes of good practice, recommendations, etc.). It was only in 2013 that soft law was included, complementary to "hard law" whose development occurred at the same time but not in competition. The Conseil d'Etat in its annual study proposes to insert these measures into a graduated chain of normativity which would go from soft law to hard law. For Jean-Marc Sauvé, vice president of the Conseil d'Etat, "there is no contradiction between the recognition of soft law as well as its expansion and better quality of law. By giving greater power of initiative to actors, and beyond more responsibilities, soft law helps oxygenate our legal order"[CON 13].

Paradoxically, it is the "EU data protection package" adopted by the European Parliament and the council on the April 27th, 2016 which will promote the evolution of the legal framework. This package consists of an (EU) regulation 2016/679 on the protection of individuals regarding the processing of personal data. Applicable to civil and commercial matters, it makes up the general framework for data protection. This regulation known under the name *Règlement pour la Protection des Données Personnelles* (Regulation for the protection of personal data) is applicable in the European Union from the May 25th, 2018. Although the regulation is directly applicable, it nevertheless provides room for maneuver, allowing member states to provide details or more guarantees for member states. Although France has made the choice not to repeal the law of January 6th, 1978 known as the Data Protection Act, it, nevertheless, made some changes to a bill submitted to parliamentarians. Through illustration, article 1 of the project entrusts the CNIL with the mission "to promote a secure legal environment through instruments of soft law whose normativity is graduated: establishment and publication of guidelines, recommendations or standards meant to facilitate compliance and preliminary risk assessment by processing managers and subcontractors, the publication of reference methodologies for the processing of health data, encouragement to develop codes of conduct etc."[2].

However, we observe, by reading the statement of motives for the bill on the protection of personal data, both a lack of definition of soft law, which is in itself a problem, but at the same time a diversity of measures, enacted according to specific needs.

Regarding this enumeration, the "instruments of soft law" present common, cumulative and constitutive characteristics of this new law. These devices seek to modify professional behaviors (similar to codes of good practice), and they do not intrinsically have any binding power (such as recommendations or standards). And finally, by these devices, professionals seize the field of lawlessness, where they establish a written law filling in the fields where hard law is inadequate. This last condition assumes that the formulation and structuring of flexible legal devices is similar to the rule of law.

They therefore fall under the regulation of a domain, without constituting a regulation. Situations regulated in this fashion often carry economic challenges with a high potential for innovation. Connected objects in healthcare already belong to the sphere of soft law and fit into, through their "technological agility", the very functions of this law.

It is therefore necessary to question their functions. These instruments can be replacement devices. They appear in the absence, in case of deficiency or the unsuitability of hard law. However, they cannot go against regulations. They appear to support the development of new uses for connected objects, carriers of risks. The collection of health data inherent to connected objects presents potential dangers regarding the respect of private life. If these risks potentially exist, they are still difficult to measure and quantify. Legal texts cannot "anticipate" *a priori* all situations linked to the use or misuse of healthcare objects, for example. However, an essential question concerns the respective responsibility of the manufacturer, and of the user who must use the healthcare objects in a precise manner. The development of a guide to good practice for connected objects and not medical devices is the illustration of this approach which aims to compensate

2 Bill on the personal data protection act. Statement of motives: http//www.assemblee-nationale.fr.

for the legislator's silence. They can also be tools for anticipating hard law, thus preparing the parliament's work. The fields of artificial intelligence and robotics are particularly affected. It then concerns an interdisciplinary dialogue with professionals to prepare the rules of hard law in advance. They then appear under their primary function, the regulation function of a specific domain for certain activities.

Developed in a transparent way (co-development, co-construction), soft law instruments attract the adhesion of interested parties. However, there is a real ambivalence to these devices, which cannot be found in hard law: are these (socially) disguised restrictive rules? Can one envisage a jurisprudential construction around responsibility, under the assumption that when a user is confronted with behavioral standards, they stray away from it? Can a judgment refer to soft law? Can a company ignore these devices? Do the rules only apply to those who follow them? There are also other psychological aspects: will the professional who doesn't follow these rules be put on the bench in their community, in society? Hard law is accompanied by legal sanctions; soft law would therefore be accompanied by social sanctions. A company that does not accept a guideline for its sector runs the risk of being potentially ostracized by society.

Beyond these questions, soft law has already entered our legal system. Through two decisions made on March 21st, 2016[3], the Conseil d'Etat (CE) accepted to appeal for annulment against acts of soft law. In support of this decision, it assumes that the contested act can produce significant effects, specifically of an economic nature or when it is intended to greatly influence the behavior of those it is meant for. Such acts, which until now were not susceptible to legal review since they had no legal power, are recognized. This law was the subject of consecration by the court's judge and by the European Union through "guidelines".

In conclusion, soft law contributes to the quality of the law. This quality is due to the confrontational way in which it takes shape. The field of connected objects in healthcare is a field of predilection, as most texts regulating this area fall under soft law.

3 CE, 21 March 2016, Société Fair vesta International GMBH and others, and CE, 21 March 2016, Société NC Numéricable.

9.3.1. *Discussion*

This analysis reinforces and legitimizes the will of actors who produced soft law without knowing it. In the field of connected healthcare objects, the term "well-being" is used to overcome legal requirements. We separate the domains, without making any links; the contract remains, which is restrictive and replaces the law.

The steps taken in this domain (guidelines, good practices) are precisely a search for the construction of soft law to occupy this field. Eventually, it will be difficult not to refer to it, and thus these measures indirectly generate restrictive rules. Ultimately, social sanction will exist.

We are aware of the difficulty involved with producing hard law, of the way in which these technologies jostle the established order. Which unconstrained commitments are associated with the individual who just bought a connected object and the supplier? There are disseminated rules with shortcomings as the law cannot keep up. Behind connected objects, there is the "quantified self": what are the issues, what are we going to do? In addition, the question evolves over time, along with technologies.

The distinction between well-being and health is evoked as well as objects referring to well-being. In reality, we are in a gray zone as there are rules of hard law to produce. There is a healthcare dimension which is burdensome, and at the same time, indeed, the time of transformation of laws is almost insurmountable. We are practically forced to be lawless. Soft law, precisely, slips into the gap left by the law and bridges it through a co-creation of rules resulting from practice. It thus anticipates the law, and avoids the "no man's land" in areas where economic activity is developing: connected "well-being" objects produce health-related data which is collected.

A shift appears between the method of development of the HAS' texts, behind closed doors, resulting from its regulatory mission, and the concerted modalities of soft law. This became clear in the case of developing guidelines for good practice of non-MD connected objects.

In soft law, authority could be limited to arbitration. At the same time, no sanction is associated with the non-application of this text, except that its non-application eventually harms other stakeholders. Therefore, this may require some form of control.

9.4. Should we consider specific economic models for connected objects?

The vision presented here is more about expertise than practical experience, even though different connected objects are designed and tested within the IMT (Institut Mines Telecom). This vision concerns the economic dimension. More specifically, the questions put forward are the following: which economic models should be used for connected objects in healthcare? What evaluation should be carried out?

Are there specific models for this type of device? In reviewing company offers, it is not clear that this is the case. The stakes are high: we recall that, according to the Gartner Group, there were 6.4 billion connected objects in the world in 2016.

9.4.1. *Economic context*

According to McKinsey [MCK 15], the annual value of connected objects in the monitoring and treatment of chronically ill patients could reach 1 trillion dollars in 2025.

In healthcare, we distinguish three types of connected objects [MCK 15]:

– objects worn on the human body or "wearables";

– implantable, injectable and ingestible objects;

– other objects ("not wearables").

The implementation of these objects would cost 171 billion dollars today: in other words, the growth of this activity should be very strong in the coming years.

The major impact of this evolution concerns patients with chronic diseases: a better quality of life for a longer life; 15 billion is currently spent worldwide on care. This expenditure is to be compared with the indicator of years of life by years spent in with a disability (DALY), an indicator that needs to be improved. The QALY (qualified adjusted life year) indicator of the number of healthy years should also be affected.

All this creates a lot of opportunities for companies and new business models to appear. New value organizations are emerging in the value chain at different locations. If it is not always directly in terms of connected objects, they are a factor of evolution in the rest of the chain.

A reference on this topic is the book written by Rifkin, *Zero Marginal Cost Society* [RIF 15], which recalls that information has zero marginal cost (an additional unit costs nothing); however, connected objects are cheap by the unit. Their diffusion is inexpensive. How will this be translated: in terms of connected objects or services? The question at hand is that of the integration of expenses in the service of quality of life, especially for working individuals. The productivity of work is not measured in this model, but it could be. The indirect costs of chronic illnesses are potentially very heavy.

9.4.2. *Business model*

According to [TEE 10], a business model is defined as follows:

> "A business model reflects management's hypothesis about what customers want, how they want it and what they will pay, and how an enterprise can organize to best meet customer needs and get paid well for doing so."

The question of economic evaluation is different from that of business models. The latter is a representation model of income structure of an activity to visualize its future development and analyze its structure and viability. An economic model describes the principles according to which an organization creates, delivers and captures value [OST 10]. These actors have published a representation that is today referenced: the "Business Model Canvas". It describes the interactions between the partners, customers and the revenue structure. It explains how to respond to what the customer wants, how to do it, who pays for what, etc.

9.4.3. *Business model of connected objects*

IMT Atlantique participated in research funded by the ANR on the theme of business models for e-health (ANR-13-SOIN-001 BBM). What can we deduce for e-health? For such an analysis, we must not limit ourselves to connected objects themselves, but must consider them in their environment. We can distinguish the following configurations, corresponding to different offers:

– the connected object alone;

– a package of connected objects and service platforms;

– a medical device.

The first case corresponds to the offer of the company, Withings (Nokia) which mainly supplies connected objects.

In the second case, the company provides a connected object and services such as monitoring and measurements. It articulates everything and offers a support service. It is a strategic choice to include or not include the object in the price of the service. The net economic results are affected by a strong investment to set up platforms and infrastructure. These companies are waiting for a return on these investments.

Finally, in health, there is the category of MD. Once again this is different, particularly due to the costs involved in complying with regulatory requirements. An example is the company Voluntis.

Among the business models, we must also distinguish the case according to the recipient of the offer: B to B (business to business) where the sale concerns a company, as is the case for the offer of Bluelinea. There can be a professional intermediary, but the consumer is affected (B to B to C). Finally, some companies address themselves directly to the consumer (B to C). These three cases are found on the market, although their exact distribution is not known.

There is a global refocusing of economic value on services, which allows a high added value. It is known as the "economics of functionality". There is less interest in the value of a good than in the value associated with the use of that good.

9.4.4. *Multiple partnerships*

In Figure 9.1, the value chain diagram identifies several actors likely to contribute to value creation: designers and manufacturers; connectivity component manufacturers; telecommunication networks; cloud; software editors; integrators; service providers and end users, etc.

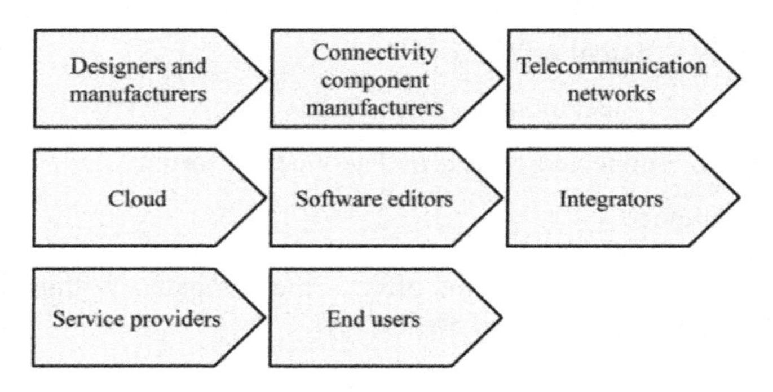

Figure 9.1. *Value chain diagram*

Because of this, the choices are complex. There is value creation everywhere. The transformations envisaged must produce useful, usable and used information for the actors of the sector, who are potentially new partners, and have new roles in the emergence of connected objects: doctors can be prescribers and health insurance is concerned with a possible refund. Mutual funds and insurances can be both financers and/or prescribers or can even participate in reimbursement.

In addition, we are witnessing the end of a risk mutualization logic with a pricing system that leans towards the personalization of care according to personal characteristics of the insured person. The pharmacist, recognized as a medical professional, can be a prescriber, but they are also a salesman of solutions. Pharmaceutical laboratories are seeking to engage themselves in the development of solutions but can also fund them.

It is necessary to pay attention to the Cloud in this chain. The Cloud is often used because it involves not investing in other means, i.e. it is cheaper. However, on the medical side, it is problematic because it is not known where the data is going. Data security is a strong requirement. Technically with the Cloud, this security is hard to guarantee. It is therefore important to question the viability of the business model. In addition, security requirements will generate costs and some position themselves as "data security actors"; the latter has the power to decide on the forwarding of data.

9.4.5. *Different economic models adapted*

Is it necessary to change economic models for connected objects? In reality, the answer is somewhat negative. There already is a great diversity of models:

– subscription pay-per-use models, such as Téléalarme;

– the "long tail" model: this offers a great number of niche products, with each of these products being sold relatively rarely;

– the model of multifaceted platforms;

– the free model: the segment that does not pay is subsidized by another component of the economic model, with three models:

- platforms (publicity) – noting that the advertising of health products is regulated, and sometimes prohibited,

- freemium: this refers to a model in which the basic service is free, and the more advanced services are paid for,

- the freebie marketing model or the "Razor/Razorblade business model" cross-subsidy or captive model: a free or inexpensive initial offer attracts consumers into the chain of repeated purchases. Examples of this are printers and ink cartridges;

– donations and crowdfunding;

– the bricks and clicks model: integrating the Internet model in the consumers' behavior in order to promote both an online (clicks) and an offline (bricks) presence in the discovery, ordering and payment of products and services.

9.4.6. *Value of data (C. Krychowski, ANR-13-SOIN-001 BBM project)*

The reflection below comes from a summary produced by C. Le Krychowski (Telecom Ecole de Management) as part of the ANR

BBM project. When we talk about connected objects, we think of the strong value of data. The potential value of data is perceived by companies, but it is currently very difficult to envisage their monetization in health. It is then still difficult to envisage a business model valuing the data generated by patients.

Patients are indeed the owners of their data, and they do not want their data to be exploited by banks or insurance companies. In addition, doctors are reluctant to compare an individual's data with general statistics, because each individual is unique.

In the absence of a "data market", we can identify three main ways in which the value of data can be exploited:

– use data as part of clinical research;

– use data internally to generate customer loyalty (the payer), by giving them aggregated statistics, to improve platform efficiency, or to partner up with actors who do not have access to the end user, and therefore show a lot of interest in data in order to better familiarize themselves with end users;

– certain companies may purchase "growth options" (financial options) based on future data monetization.

9.4.7. *Link between the economic model and evaluation*

It is possible to establish a link between economic models and the expected level of evaluation evidence. Indeed, Figure 9.2 shows the type of financing available depending on the positioning in a market, from public to private, and the type of evaluation expected.

Figure 9.2 shows how, depending on the solution, funding can be secured: by the Public Power (left axis), by the market (right axis) or by a combination of the two (between the two axes), in a variable way and by different organizations, depending on the value propositions made.

This involves measuring this value and defining the conditions for this evaluation. For example, in the case of telemedicine, a medical act designed to be reimbursed, evaluation is undertaken via the realization of clinical trials. However, there are also medical devices on the market (blood pressure monitors, etc.) for which the consumer is willing to pay. In the middle, next to health insurance, other funding can intervene, for example if the solution has social value (people with loss of autonomy, those who are disabled, etc.).

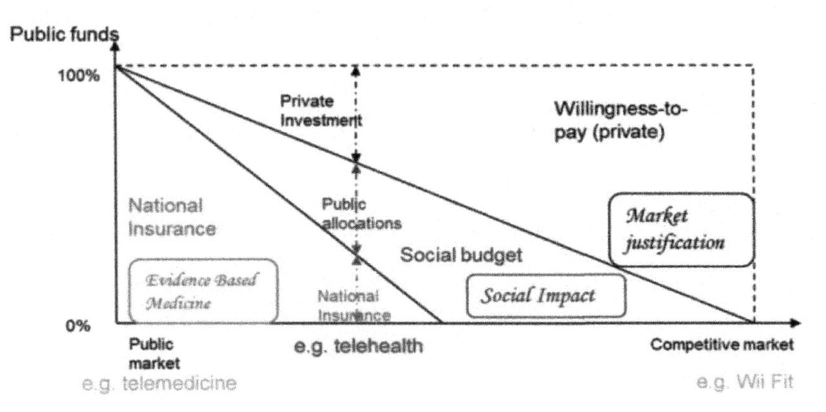

Figure 9.2. *Expected level of evidence depending on the market concerned (source: Le Goff-Pronost and Picard, [LEG 11]). For a color version of this figure, see www.iste.co.uk/picard/value.zip*

As soon as we leave the right axis, we can legitimately ask ourselves the question about the interest in new forms of evaluation and some reflections exist on this topic. Therefore, for example, within the OIIS project, *Ocean Indien Innovation Santé* (Réunion Island's TSN program), work is being conducted to try and evaluate the program's social impact. Some of the implemented solutions include connected objects and so will be evaluated. We are at the beginning of this renewal of reflection in terms of evaluation: for the time being, evaluation in health by the Public Power is dominated by clinical trials.

9.4.8. *Reimbursement of connected objects*

Reimbursement is possible if the connected object claims to be a medical device (CE marking).

For a reimbursement by the social security, the device must be registered on the LPPR (list of products and services reimbursed) as a medical device for individual use. This list of devices includes: glucose monitors, electrodes, strip and sensors, injection pens, instruments for measuring clotting, continuous positive airway pressure (CPAP) and advanced flow meters.

According to the *Conseil de l'ordre des Médecins* (French Medical board) [CNO 15], reimbursement is possible "as soon as the evaluation of the connected objects effectively recognize its benefits on individual and collective health".

Reimbursement can also be proposed by mutual funds. This is typically an annual fee to support the purchase of small connected medical devices.

There are different levels of evaluation for reimbursement. The evaluation of the proven benefit can be aimed at three different dimensions:

– medical: the medical service provided, improving the medical service provided;

– medico-economic;

– quality of life.

The evaluation that is referenced today is that of clinical trials (randomized, observational, etc.). However, the impact of prevention (monitoring indicators) is somewhat difficult to measure.

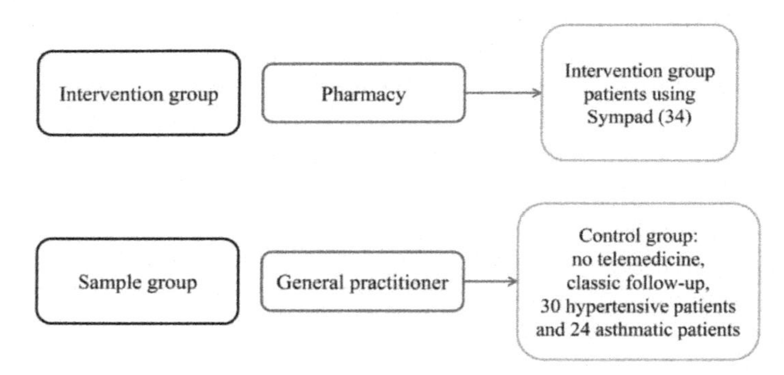

Figure 9.3. *Evaluation protocol of Project Sympad*

Figure 9.3 presents an example of a project's evaluation protocol involving connected objects (Project Sympad – E-health 1). Project Sympad involves the pharmacist in terms of connected objects: they follow up, advise the patient, and invite them to measure their blood pressure and their weight. A medico-economic, observational, multicentric study of remote monitoring of patients in pharmacies was thus set up. The study lasted 16 months, of which 10 involved recruiting and 6 of patient follow-up. The project concerned two diseases: high blood pressure and asthma. A cost-effective study was proposed with the main criterion being the evolution of high blood pressure and asthma and the evolution of forced expiratory volume in one second (FEV1).

The economic evaluation of this project failed due to a lack of data and a sufficient use of available connected objects in pharmacies by the recruited patients. This solution was probably proposed too soon. The patients attended the first measurement, and most of them did not attend the second one, even though they were supposed to come back after 3 months and then again after 6 months. Only one person followed the protocol to the end and the expected patient–pharmacist relationship did not function as expected. This raises the question of patient ownership and the acceptability of such objects in a context outside of their home.

9.4.9. *Discussion*

The figure of $171 billion representing the value of connected objects can be surprising, especially since in recent studies this would concern 60 or 80% of "wellness" solutions. People purchase physical objects in order to *use* them, so it is not the same thinking as for drugs. Usually in health, these devices are equated with drugs. Health insurance plans to devote 0.3% of its budget to the reimbursement of these objects while using the existing model for drugs but this is not well-adapted. We must find a "fair price", but we do not really know what that means and it is very complicated. Health education and patient education are concerned by these objects, but it is currently not very developed. Creating a business model by taking into account the impact on population health is difficult and evaluation is very complicated.

We can evoke, for example, the applications that count the number of footsteps we take in a day. What is the ultimate impact on the health of this tool's users? It is therefore necessary to mention added value before mentioning a business model.

Project Sympad highlights the issue of the provision of connected objects by a third party (often to avoid purchase by the patients themselves). This mirrors a project carried out in a neonatal department in the city of Nantes. Parents and children are brought in, by showing them "eye-catching" things to make them want to come back. With pharmacists, another dimension comes into play: the time they had to devote to this new relationship in an already limited amount of time. The remuneration for this follow-up time, which could also be considered as time devoted to prevention, could be envisaged.

Another example cited is that of the connected pillbox: this object contributes to controlling the overall cost of oral care. The packaging is not suitable for the city. If a box contains 30 drugs and the patient must stop taking them after 2 days, the rest are useless. There are intakes in cycles, and at the end of the cycle, what is not taken is lost. This represents considerable costs. In oncology, treatment costs vary between €4,000 and €13,000 per month in France. It is understood that if we discard treatments corresponding to 20 days, the cost is considerably high. It is also necessary to simplify the intake of drugs, with a simple, ergonomic solution facilitating adhesion. Finally, it is necessary to reduce the situations that require urgent re-hospitalization.

Connected objects of Santé Landes (Section 4.3.2.1) raise another issue: the implementation of these devices (connected watches and ePad) was expensive. To obtain a satisfactory functioning, a person often had to come from Santé Landes or the center of Bergognie and go to the patient, sometimes up to 3 or 4 times. In the month and a half that followed the first installation, it was shown that this can cumulatively represent up to 2 days/person.

9.4.10. *Conclusion*

The economic models and evaluation methods are linked. Will new evaluation methods appear? It should be noted that the "references and guides for good practices"[4] do not produce the HAS on mobile applications and connected objects include families of indicators that go beyond classic criteria and resemble the axes of the multidisciplinary GEMSA grid published in 2011 [LEG 11].

---

4 https://www.has-sante.fr/portail/jcms/c_2681915/fr/referentiel-de-bonnes-pratiques-sur-les-applications-et-les-objets-connectes-en-sante-mobile-health-ou-mhealth (see section 8.1.4).

10

The Question of Technique

At the technical level, a connected device is composed of one or several sensors, various systems for converting received signals, all connected to a wired or wireless transmission system, which is controlled by an application software (typically mobile or web service) to transmit information via a mobile application.

In Part 1, we saw how these connected medical devices are likely to transform the experience of users and professionals as well as the healthcare pathway, in living spaces or in the case of hospitalization, from admittance into hospital until the return journey home.

To be effective, this type of system which multiplies the data collection capacity must also allow the data exploitation process, and the visualization and sharing of this synthetized data ("Big Data"), as discussed in Chapter 6. Artificial intelligence is in fact now recognized as a factor of scientific progress, especially in the healthcare sector [VIL 18]. However, many technical challenges still need to be overcome regarding measurements, data collection and processing. This particularly concerns the control of security, availability, integrity, confidentiality and data interoperability requirements.

In this chapter, from field examples, we will first illustrate certain difficulties and risks encountered when mobilizing connected objects or mobile applications. Second, we will present several answers to be considered, by the mobilization of a "systems engineering" type approach. The goal of this chapter is indeed to make the reader aware of the reality and

---

Chapter written by Denis ABRAHAM and Robert PICARD, with contributions by Henri NOAT.

nature of the risks posed by connected technologies and the need to mobilize useful expertise to avoid them.

Regarding the methods and tools allowing the control of these risks linked to the use of connected medical devices, we refer the reader to the book "Connected Objects in Health: Risks, Uses and Perspectives" [BEY 17].

10.1. Feedback

The following elements regroup a certain number of difficulties encountered in various projects and whose origin was attributed to technologies: connection problems, network operation, maintenance of solutions and inadequacy of technical trade devices. These aspects are dealt with successively in the first section of this chapter.

10.1.1. *Connectivity*

There are different connection protocols through which connected healthcare can be implemented. These protocols differ in their costs, the fact that they are published, their diffusion and technically their security and reliability, as well as the constraints they place on the user (whether or not the user must manually activate the link, electric autonomy/need to charge the object). Reliability and safety in terms of functioning are also dependent on the amount of connected equipment in a solution, especially when they come from different manufacturers.

Connections can be managed using an interface application supplied by the manufacturer: this solution works well when there is only one sensor or several sensors from the same manufacturer. But in the case of a heterogeneous solution integrating several connected devices it is necessary to cope with different applications associated with a variety of user interfaces. The interaction with the overall system becomes complex and cumbersome, and the design, which is strongly constrained, cannot be optimized. In addition, data flows often escape the designer, since manufacturers privately collect this data to value it according to their business model.

Connections can also be managed by an application developed by a designer, based on the manufacturer's toolbox or "SDK" (Software Development Kit), which allows for direct connections with concerned objects in a personalized way. However, medical procedures are not necessarily in a position to cope with all types of devices (e.g. different types and versions of device operating systems). Manufacturer software updates must be carefully managed; otherwise, a high level of risk regarding dysfunction will be anticipated.

10.1.2. *Network engineering*

Group feedback shows that there is still a need to improve the overall performance of systems based on connected objects. There are still too many difficulties encountered when implementing the products available on the market. Consumer technologies are often unsuited to the requirements of healthcare (volume, frequency, reliability) that appears as a specific domain.

In an ideal world, manufacturers should increase the quality of the solutions they offer, and progress is also expected regarding transparency (data protection in proprietary applications). Regulatory evolutions go in this direction (see Chapter 9). Requirements relating to flows of volume and quality must be met in order to provide the service.

There is a particular problem associated with real-time monitoring. In the case of a continuous use of connected objects, the volume of data can be problematic: for connected objects using “BLE” (Bluetooth Low Energy), which tends to impose itself as a standard, it is necessary to consider the fact that the transfer rate is limited to relatively low value. Owing to this, the transfer time is too long to use this type of object in many healthcare applications. This is the case, for example, with connected patch-type objects processing multiple forms of data or sensors associated with electrocardiograms producing complex data. For these objects, energy consumption is an additional challenge.

In this context, the mobilization of good skills, those of network engineering, is a point of vigilance. Relatively sharp technical skills are required to clarify and mitigate the vulnerabilities of the service set up at network level to characterize it, so that it meets the required needs.

It is important to discuss the notion of “real time” which is defined differently according to the point of view taken. Real time corresponding to physiological parameters is generally satisfied with time constants of the order of seconds, or more. However, in certain electronic transfers, the requirements can be much higher.

There are industrial solutions where the communication protocol is adapted specifically to the sensor. What guides the choice of protocol is the need to achieve a good level of security, for the volume of information to be transmitted, and to have a fully automated system without human intervention, or not. Power consumption must also be studied because a sensor must have an operating autonomy compatible with its use. Sensor data must be secured, and it varies from sensor to sensor. In addition, it is necessary to provide opening possibilities to connect sensors from other manufacturers. To achieve these requirements, the principle is to rely on a recognized protocol, and to place above an open bus. It is best in this context not to rely on the manufacturer's maintenance process.

An example of sensor networking is presented below.

10.1.2.1. *A sensor networking approach*

This presentation displays the developmental work carried out within the company Sesin to configure the Hadagio solution in the patient's home. This is the preliminary study of the solution, apart from any study on use at this stage. It consists, from the needs to be fulfilled, of making choices regarding equipment and associated modes of transmission.

The solution's specifications had to consider the following constraints: constraints of cost, space, computation power, energy consumption, operational dependability, opening to sensors other than those integrated by the company (open system) and of security. These requirements are very different from what is conventionally displayed for products on the market. They relate to the operation of technical devices, and in particular their capacity to interconnect with a remote server.

Regarding this, the first task consisted of positioning the various known protocols according to their range (maximum distance of two connected pieces of equipment). The retained distance was 100 to 120 m, ensuring service to all rooms of any apartment. Only one protocol satisfying this criterion was nevertheless eliminated, because of its high operational cost (Sigfox). There remained four possible protocols: Zigbee, Bluetooth, Wi-Fi and RFID.

A preliminary analysis concerned the documentation of these protocols according to the following criteria: IEEE reference standard, memory requirements, battery autonomy, number of nodes, transfer speed, range, operating license (paid or not), popularity, presence of encryption, complexity and frequency range.

This analysis was carried out by means of several tests.

At this stage, the two preselected solutions have been Zigbee and Bluetooth. Zigbee has the advantage of being compatible with the IEEE 11073 medical equipment standard, for connecting a pacemaker, for example.

The tests focused on a MESH-type configuration. This mesh network architecture means all equipment (sensors, machines, concentrators, etc.) are peer-to-peer connected without central hierarchy, thus forming a "net" like structure (known as a mesh). Owing to this, each piece of equipment receives, sends and relays information. The advantage of this architecture is that if part of the network fails – or if there are particular propagation difficulties caused by the building's configuration (concrete walls, etc.), this part is not isolated because data is reported by other nodes. The security relies on the IEEE 802.15.4 protocol; security is provided by a strong encryption: AES-128 bits. For authentication, only declared nodes can connect to the network. Other advantages and disadvantages were revealed by these tests: transmission was sometimes disturbed and there was a need to replace the energy source – which led to providing a signal to the user when the battery must be changed.

10.1.3. *Maintenance and management of change*

Ideally, it would be desirable to have a formalized and published maintenance process (SDK) in relation to the evolution of iOS/Android. But in practice, the changes arrive inadvertently and the number of users of global solutions is far too high to reach the "ideal". Apple or Microsoft cannot be held responsible for each disruption to the functioning of their system with each new version. Without being able to impose a provisional schedule on these global players, it has been suggested that they collaborate with specialized professionals, in order to inform themselves as soon as possible of this type of change and to anticipate problems in order to better solve them.

10.1.4. *Autonomy, reliability and adoption*

For sensors measuring blood pressure, or blood glucose, the batteries should last three years. Energy consumption must therefore be limited. Architecture is thus a function of both the intended usage and control needs. If the flows are interrupted, we must know how to react. This is part of the system's architecture. Similarly, data protection must be ensured so that it does not end up somewhere on a Cloud. The architecture depends on the uses.

Connected objects have a strong attractive and appealing effect on users, who are both patients and healthcare professionals. One difficulty is that of channeling this appeal and advancing step by step. Indeed, connected objects are not necessarily developed for the populations solicited in medical programs: they are more or less technophilic and ready to adopt the technologies; this is also affected by their age. We would like to witness more intuitive ergonomics and for usage evaluations to always be carried out on this type of population.

It is necessary to develop the user experience and the uses. This is a recommendation which should be offered to manufacturers who want to promote connected objects in healthcare. There are prerequisites: it is necessary to start with the cases of use, considering very diversified populations.

10.1.5. *Conclusion: technological mastery, myth or reality?*

10.1.5.1. *Real difficulties*

In real life, everything does not work very well at this stage. One experiment reported that in the group only 30% of the patients involved in a cohort continued using their sensor after one day. More worrying still, most measurements taken by a nurse in parallel to the tests correlated well with the sensor's measurements, but for heart rate, there was not really any correlation. However, abnormal cases (only one in this case, during the trials, in the cohort in question) were correctly identified by the patch. Thus, mastering this type of technology still involves certain difficulties. In one of the presented cases, tests carried out over several months, in patients without any initial training, showed that the sensors could perfectly integrate into the patients' daily lives. However, the sensors which were not important for controlling health status were only supported for two or three weeks, while the glucometer for a diabetic patient was not rejected.

Another problem concerns the reconciliation of measurements coming from solutions provided by different companies, collecting data according to different principles. These same companies may also wish to develop their solution. In a context where the regulatory requirements are increased, where the application software itself acquires medical device status, changing the algorithm will require a recertification of the product. This will be an obstacle to innovation, as well as meaning additional costs and delays.

The sensors are sometimes badly adapted to real patients. What works very well on mannequins in a laboratory can be revealed as inapplicable in real situations. It is a little-known statistic that 70% of new "innovative" products and services from the industry do not develop a market; they are not adapted, or are misunderstood by the recipient "client".

Many providers seem to place themselves on the market with devices that ultimately appear to be gimmicks, due to a lack of prior knowledge of problems specific to healthcare; there is no "feedback" from previous offers.

10.1.5.2. *The need to evaluate the value of technical products and services*

It is therefore necessary to be aware of (understand, precisely describe) the offered functionalities/services, the operation limits, the types of malfunction and the conditions of dysfunction (type of indication/contraindication with adverse effects).

It would consist of following an AMM (drug) type protocol, a procedure that makes it obligatory to specify these points as is done for drugs.

Two questions then arise:

– Can we wait 15 years before using new technological solutions because of these increased constraints? The answer is clearly no; there is no question of waiting so long to benefit from what is technologically feasible.

– But should/can we use them "with our eyes closed" (without knowing/understanding them)? This is clearly not an option. The risk is that the technology does not function when it is needed: for example the alarm button at the time of a fall, because it is not accessible, not waterproof in the shower, etc. Service providers have highlighted a high percentage of “false alarms” corresponding to the satisfaction of a real need: to break loneliness by talking to a kind person.

What is the answer to this question? To validate what technology can really do and its limits: it is the validation/qualification (functionalities, offered services, limits, malfunctions) component of Living Labs and Test Beds[1].

10.2. System engineering as an answer

The question of measurement has already been mentioned in Chapter 4. Measurement is required to monitor physical parameters, and also to collect other information: environmental information such as meteorology, ambient temperature, hygrometry and air quality, or domotic-type information such as that provided by motion sensors – in different living areas – or associated with actuators: opening doors and windows, opening or closing shutters, etc.

At the technical level, measurement not only refers to the notion of sensors and actuators, which are important subjects, already mentioned and repeated below, but also calls for a notion of a complete system. Such a system can be segmented as follows:

– sensors;

– collection network(s) (for information from different sensors, most often close or relating to the same person);

– network(s) transporting measured information;

– storage areas;

– algorithms: exploitation of raw measurements, to design a value-added service;

– identification of the user or of the value-added service(s);

– tools and systems to access value-added service(s) (including capture where appropriate)[2].

1 For more on the topic of “Test Beds”, see section 8.3.

2 For more on the comparative advantages of captured versus collected data, see Chapter 5.

Limiting the technical problem to a component-by-component approach does not allow any designer to consider the extremely important notions referring to the characteristics of a complete system. As such, we will mention:

– end-to-end data security;

– ethics in the technical field. Indeed, the notion of ethics in the sense of "doing well" is generally limited to the medical, medico-social, healthcare and social components. But from the technical point of view, it refers to the notion of quality: "the quality of a technical chain relies on that of the weakest link". This ethical aspect of connected healthcare devices and systems should not be forgotten;

– the service's resilience, its ability to resist undesirable events and remain available, possibly in a degraded way. This depends on the nature of value-added services and how they are used;

– among these events, the risk of a cyberattack is intrinsic to any connected system, including connected objects (the Internet of Things).

We therefore need to think that end-to-end systems engineering, in accordance with the state of the art, does not limit the analysis to only those aspects corresponding to technical challenges or domains that are sensitive to users or sponsors, or even in the field in which the project leader has skills. No link in any technical-economic chain should be ignored. Any partial approach would not only be reductive, but above all, behind "blurred lines", could obscure the weaknesses of the overall system which could prove disastrous.

Indeed, the implementation of a faulty "connected health" system, even for a single weak link, can lead to strong disappointment among users, beneficiaries and clients engaged in this type of solution. As a result, it will be very difficult to reinitiate a relationship built on trust, a service, an activity and a sustainable business.

10.2.1. *Sensors*

Concerning sensors, it is first important to listen to the medical body to collect what seems to be appropriate to measure and monitor in terms of parameters, whether it is of a physiological nature or linked with individual living environments (this may be closer to devices such as automated sensors or actuators) for which benevolence is considered.

This benevolence is reflected most often with a risk of loss of autonomy related to a disability, aging, chronic disease, a difficult treatment or hospital discharge – especially in the case of ambulatory care or when the duration of hospitalization is minimized/optimized and a follow-up 48 hours after discharge from hospital is of greater importance to better manage post-intervention risks.

Once the relevant parameters are identified, it is first of all necessary to identify the possible existence on the market of sensors that are likely to be able measure them. In this case, it is necessary to characterize them for their performance in terms of metrology (accuracy, fidelity, sensitivity, resolution/precision and measurement range) in order to establish their use value.

More often than not, sensors exist. However, in cases where sensors do not exist or when those that do exist do not present the required qualities (to be compared with the needs expressed regarding the envisaged uses of measurement), then it is appropriate to turn to research. The goal is to design transducers capable of converting a physical or chemical phenomenon into a different measurement, for example electric, which can be easier to manipulate and characterized in terms of metrology, always with the goal of answering the needs expressed by the medical body.

10.2.2. *Collection network(s)*

This is a network capable of handling information originating from different sensors, often related to the same person.

In general, when someone plans to be benevolent towards a third party in terms of the loss of autonomy, even partial or temporary, we are led to simultaneously look at several parameters in order to analyze the potential concomitant evolution, interactions, correlations and so on.

In this case, it is necessary to set up a very local network, at the scale of the body, a room or a house, capable of facing situations of mobility or on the contrary, stationary situations, capable of collecting data from devices relating to the same person and routing this data to a tool that harbors all of this data, whether it is related to parameters with a physiological connotation or environmental data that can be related to that of home automation. There exist several types of networks that are capable of meeting this need of collection, wired or wireless. In this second case, several frequencies, e.g. types of modulation, are identified according to the needs of use (distance, noise level, ambient environment, etc.), different protocols (a kind of vocabulary used to structure the communication between transmitters and receivers and to give it an efficiency as much concerning the robustness of communication even in noisy environments as the security of the information or the multiplicity) and the possible heterogeneity of interlocutors, etc.

The functionalities related to the collection network are often implemented in so-called "hubs". Among the hubs, we can find smartphones and depending on the technological choices made in terms of collection networks, all the necessary ingredients for collection networks. However, smartphones, if they are likely to possess the required qualities (among others), can be considered as expensive compared to a dedicated hub and have a certain heterogeneity/diversity that should be properly managed (exploitation system, versions, non-controllable changes imposed by the equipment manufacturers, necessity to pass by "blinds" for certain operations, etc.). In this case, if the individual benefitting from the benevolent service possesses a smartphone, it is tempting to rely on it (its cost is thus covered for the other service for which its acquisition has been already planned), with its advantages and its disadvantages (that must not be ignored/forgotten).

Standards such as Smart BAN, Bluetooth, Bluetooth Low Energy, Lo-Ra (Long-Range) WAN, Wi-Fi, as well as the so-called "telecom" standards such as GPRS, 3G, 4G and soon 5G, are quite plausible and should be considered in terms of their technical and economic characteristics, as well as their advantages and disadvantages.

10.2.3. *Network(s) for transporting measured information*

The previous section described network engineering, or at least one that applies to the very local data collection network.

The same approach needs to be applied concerning the data transport network, which will probably include several network technologies: wireless, wired, private network, public network and Internet. This network, in a broad sense, must not only consider much greater distances and increased risks, but also a more constant flow of information, potentially including information on a larger number of individuals (which can be personal information that could have real medical connotations), and more ambitious geographical perimeters – these networks usually do not recognize borders but are likely to need to take into account more or less homogeneous and continental national legislations.

10.2.4. *Storage area*

There is a need to store the measurements at different levels of system architecture.

10.2.4.1. *In the sensor:*

Beyond the invitation to measure such and such parameters, the question of the periodicity of the measurement is very quickly raised. Indeed, while some magnitudes are intended to rapidly evolve (e.g. heart rate), others will see their evolution come about more slowly (e.g. weight), and it is therefore appropriate to adapt the periodicity of measurements to this notion of upgradeability.

The question of the use value of measured information also arises very quickly: in some cases, it is the instantaneous information that is expected (e.g. the detection of a fall). In other cases, the use value relates to a notion of an accumulated number over a period of time (e.g. actimetry).

Depending on the case, it consists of:

– storing information; the question that arises is whether to establish a system strategy, i.e. to decide where the computing capacity will be mobilized:

– adding a small storage capacity to the sensors themselves;

– deciding to immediately transmit information in order to mobilize remote computing, for example on the site of a manufacturer or a service operator (in the Cloud, perhaps);

– a hybrid model which is often relevant for optimization purposes;

– carrying out information processing to transform it and give it its use value (average, accumulated sum, minimum/or maximum value, peak, integral, trend, etc.). For this, it is necessary to have a computing capacity. Similar to the previous point, it is necessary to establish a system strategy concerning computing capacity. Thus, the sensor can be provided with a sufficient computing capacity to make a first/pre-processing of information;

– transmitting raw information to a remote site, well equipped with computing power to process information.

10.2.4.2. *At the level of intermediate equipment: hub/gateway*

In the system strategy mentioned in the two points above, we often have to wonder about the sensor's capabilities, beyond its qualities in terms of metrology attributed to the sensor itself. Should it be provided with a computing capacity, or a storage capacity? If so, how much?

These choices are also not trivial in terms of its size, price, and especially its energy consumption and energy autonomy, although this is not the subject.

The strategic choice, to limit it to its simplest "sensor", the device that performs the actual measurements, is accompanied by relocation of both storage and computing power to a site (e.g. the supplier's site). This choice of centralization is not necessarily a bad choice as it introduces a very strong constraint on this site.

An alternative to these two options of "all-distributed" or "all-centralized" is to provide intermediate equipment, sometimes called a "hub", which is capable of local measurements, for example that which concerns one and the same person, and to contain part of the computing and storage capacity needed for value creation. This extremely common approach leads to a star-like (hybrid) architecture that can optimize several aspects of systems engineering.

10.2.4.3. *At the central level: establishment of a database*

Whatever choices are made (see the previous section), everything leads to the constitution of a database (which can be raw or have already been the subject of certain calculations or algorithms), which aims to derive a larger use value. This database is particularly rich and contains, if not medical information, at least personal information.

This database can quickly become large, either because it includes the data of many people or because it contains information that is periodically measured, and as time passes, the quantity of measured information accumulates and ends up constituting what is often called "Big Data".

10.2.4.4. *In the restitution (and input) terminal*

Whatever its form, this type of equipment is also likely to include storage and computational capacities, which can be mobilized.

10.2.5. *Algorithms*[3]

Another major challenge is based on algorithms and calculations that can take control of the exploitation of more or less raw data in order to constitute value-added services.

This stage is key to the creation of use value for users themselves, and also for the individuals who wish to be benevolent from a distance. It then constitutes a key factor of success both scientifically, medically, medico-socially and from a business point of view. Special attention should be awarded to the site that hosts the database, constituting an essential element of the system architecture, in terms of security, reliability, resilience, etc.[4].

10.2.6. *Identification of the user of the value-added service(s)*

When designing value-added services, it is necessary to identify the user(s) of the service(s). They can be:

– the sensor carrier's shareholder themselves;

– their entourage;

3 The challenges of algorithms are developed in Chapter 6.
4 Tools such as TERALAB (Big Data platform of the Mines Telecom Institute) are extremely valuable for success in this area.

– people who would be benevolent to them while remaining at a distance;

– family caregivers;

– professional caregivers;

– doctors, etc.

For each of these profiles, it is necessary to provide and prepare the service according to the derived use, in a more or less synthetic and synchronized fashion, or on the contrary "on demand". It will be necessary to properly identify the individual to whom the information is delivered, according to authorizations to which they can legitimately claim.

Many questions must be asked and clearly stated while taking into account the notion of ethics.

10.2.7. *Tools and systems for access to value-added service(s)*

Once the user(s) are identified (see the previous section), it is necessary to design the device that will allow them to access the value-added service(s).

This can be a computer terminal, a PC, a tablet or a smartphone depending on what is readily available, and also depending on the required characteristics for the service (also taking into account the choices in terms of system architecture: centralized, distributed or hybrid, with different levels of hybridization).

10.3. Back to systemic issues

To be focused only partially on the problem, deprives us of a "vision" because this prevents us from taking into account extremely important notions such as security, reliability, ethics, legal dispositions concerning personal data, human rights[5], etc. We illustrate this point regarding system security below.

5 On the regulatory and legal aspects, see Chapter 9.

10.3.1. *End-to-end data security*

This is an extremely important aspect; any negligence on this point would expose users to risks to which they were not exposed before, using the tools in question. Any professional likely to offer such tools should naturally have been informed of these risks.

Thus, it is necessary to secure the entire chain, as presented above. Any weakness in terms of security could lead to the destruction of value and the loss of business expectations.

10.3.2. *Resilience*

In the same vein, depending on the nature of value-added services and the use that is derived from them, it is important to consider the resilience of the service (availability of services). For this, it is necessary to revisit system architecture to strengthen and to make it more robust, in order to allow it to maintain the services it provides despite the appearance of certain technical problems.

10.4. Conclusion

The presentation of the system approach in this chapter should enlighten the reader by allowing them to avoid reasoning that is too simple such as the following: "it is enough" to connect such devices (which have a given functionality) to such others… and "it works"! The need for professionalism and vigilance is mandatory for all technical issues, and specifically for the following issues:

– What paths does the information take?

– Who can access the data, and do they use it for good or misuse it?

– Who are the actors involved in the system's security? Who is responsible?

– Who ensures its proper functioning (by defining the available functionalities, the type of malfunctions that may occur especially when we leave the operating areas for which the devices were designed, the features that are not proposed, etc.)?

– What responsibilities are engaged in certain aspects?

– What are the risks, including those concerning the business plan and the ethical plan?

These topics are essential, and a certain number of them are regulated: it consists of giving them the attention necessary, in relation to their importance with regard to each other as well as any underlying risks.

This system approach is sometimes perceived as difficult and arduous to implement. However, experience shows (see section 10.1) that an incomplete vision generally results in unsatisfactory or even unsustainable solutions. The user is surrounded by different heterogeneous sensors, in terms of connectivity; they are also requested to pay several network subscriptions, e.g. multiple SIM cards for public mobile telephone networks. This leads to a very great deal of complexity due to compatibility and interoperability problems (we will thus look into interoperability before considering any notion of quality of service rendered), and reconciliation of information measured by each of them on the same consultation terminal (without mentioning the quasi-global dissemination of personal information).

It is necessary to establish a relationship of trust with connected devices, services and systems based on positive user experiences and technologies that work. The economic activity and availability of adapted responses will develop for the better if the population trusts in the viability of these tools. Thus, use value in all its dimensions, including those more directly related to technological qualities, is called upon to play an important role in the scheme that can be likened to a "virtuous circle".

It seems inevitable that efforts should be implemented to "characterize" the devices, services and systems, in order to allow the establishment of this relationship of trust concerning the devices for the patients (who carry the sensors); their entourage; their potential family caregivers, volunteers or professionals; potential prescribers; co-funders (whether they are public or private); etc.

The Living Labs and "Test Beds" (section 8.3) are called upon to play a role in this context, as a place of favored exchange between all stakeholders, and reassurance on the value propositions dependent on technologies.

PART 4

Perspectives

Introduction to Part 4

Connected healthcare is just beginning. In the introduction to this book, emphasis was placed on the human problems of appropriation and the domestication of this new approach to healthcare. Organizational issues and the diffusion of knowledge were also discussed.

In addition, the following were also mentioned: the variable reliability of connected objects; interconnection methods; the cost of setting them up; the instability of regulatory framework and the complexity of economic models allowing sustainable solutions.

In this part, we would like to shed light on certain changes to come, probably in the longer term, but which the academic world is already concerned about.

Chapter 11 develops two new perspectives on public healthcare: the first focuses on new ways of monitoring public health and evaluating therapeutic solutions by using connected devices and by exploiting available clinical data. The second focuses on prevention and the necessary conditions for the efficient mobilization of new technological tools.

Chapter 12 opens perspectives based on the introduction of practices developed elsewhere in the field of healthcare knowledge, and whose adaptation to the healthcare sector constitutes both a very important issue and challenge. The mobilized domains are design, management science/innovation management and engineering.

11

Public Health Perspectives

The perspectives studied in this chapter develop in two complementary directions:

The first concerns our ability to systematize the exploitation of data coming from clinical trials and everyday life for researching and designing new solutions. This approach, made possible by the digital world, is likely to revolutionize clinical trials, as well as upstream research and downstream prevention and health surveillance (pharmaco- and materio-vigilance with increased identification capacities prior to secondary trials).

The second underlines the need, in this context, to better understand and master the way in which the introduction of mobile applications and connected objects is likely to modify the behavior of individuals and improve the population's health. This is particularly true in the field of prevention, which will be questioned more in depth in the chapter devoted to this subject.

11.1. Clinical research based on real data

This section is a free reproduction of comments made by Professor Ravaud[1] during a working group hearing by the strategic committee of the health sector on the theme: problems posed by the evaluation of m-health solutions – September 26th, 2016.

Chapter written by Olivier AROMATARIO, Linda CAMBON and Robert PICARD.

1 Professor Philippe Ravaud is an expert methodologist in evaluating non-pharmacological treatments and interventions to improve care. In practice, he is requested by start-ups to evaluate various connected tools.

11.1.1. *Context*

The problems posed by m-health are those of its impact. It is a challenge for methodologists specializing in the evaluation of non-pharmaceutical treatments. According to the EU[2], there are now 97,000 m-health apps, with a wide variety of uses. Of these, 70% concern the general public, including "well-being" and 30% target healthcare professionals. The estimated market was worth 17.6 billion euros in 2017.

Functional evolution goes from self-measurement, the initial domain, to diagnosis, then to treatment and prevention, ultimately leading to global applications.

11.1.2. *Problems*

The number – compared to drugs, less than 50 new ones per year – diversity (permanent changes or improvements, for example, for an algorithm, multiple versions, software updates), instability and rapid obsolescence are problematic.

These applications are often integrated into complex interventions with many other components, and the entire intervention must be assessed: for example, what data is produced? What is the organizational impact?. They can generate data constituting evaluation criteria. These solutions will sometimes profoundly challenge the conventional organization of care. The applications concern: the diagnosis, screening, prevention, monitoring, prediction and modification of care organization.

11.1.3. *The issue of evaluation*

Evaluation methods must be adapted for various purposes (source: the m-health evidence workshop):

– in the development phase (tests done before/after; test 1 of N; time-interrupted series);

– tests associated with mature interventions;

2 https://ec.europa.eu/digital-single-market/en/news/green-paper-mobile-health-mhealth.

– randomized controlled trials (RCTs); continuous regression tests; extensive randomized trials.

Beyond randomized controlled trials, alternative methods are emerging.

A study was published in the journal "Beyond the randomized controlled trial: a review of alternatives in mhealth clinical trial methods" [PHA 16] concerning 71 trials conducted between 2014 and 2015, of which 80% were randomized. This work highlights the large diversity of evaluation methods, depending on the issues to be solved. There are simple questions (like sleep monitoring) and more complex ones where we wish to measure the impact of an application on a process (e.g. physical activity) or on an outcome (like the improvement of a condition).

11.1.4. *Connected objects and their impact on behaviors*

The question of a smartphone's "accuracy" can be illustrated by measuring the number of steps, for example 1,500 steps (corresponding to a desired result). The application is a facilitator of the physical activity of walking, but not a driver. However, it is the engagement strategy that matters; it is not just a case of tools.

11.1.5. *Illustrations*

IDEA[3] is a randomized controlled trial. It concerns the comparison between a consulting service (by telephone or SMS) and a connected tool associated with a web interface, for overweight individuals. The result is that the connected tool does not perform as well in terms of weight loss than conventional support!

But there are also mobile applications that work, for example the MoovCare application (see section 6.4.1).

A conclusion can be drawn in that it is impossible to use the classical evaluation framework, used for drugs, which extends over 10 years, from its initial idea to the drug being released. This approach is expensive: 100,000 to 100 million euros per study and this is not necessarily suitable for

3 https://www.ncbi.nlm.nih.gov/pubmed/27654602.

evaluating connected objects. There is a need to change the ecosystem and mobilize good methodologists.

11.1.6. *The future of clinical trials*

It is therefore necessary to reinvent clinical trials for objects developed rapidly and to reinvent systems intended to produce "evidence". Disruptive approaches are needed for rapid development at relatively low costs. Must we abandon randomized trials? The amount of data must be considered, and the cost of testing reduced.

The future will probably not be based on randomized trials. Timely data will need to be used, from experimental to observational studies. These studies must be able to "mimic" randomized trials while being shorter and cheaper. We must be able to approach the effects of treatments in this way, more realistically, and be closer to real life and develop an approach to causality from observational data. It is an unrealistic expectation that we would have randomized trials for all interventions or combinations of interventions in all patient subgroups. We need timely demonstrative evidence and the time to conduct randomized trials. Consequently, 85% of data proving the comparative effectiveness of research will come from non-experimental studies.

This forces us to think differently about research. We must think "low cost" while maintaining the highest possible level of quality, forgetting conventional methods (or rather keeping necessary or crucial conventional methods and removing the rest). It will be necessary to adopt ethical rules that are more closely linked to the evolution of techniques and with what is done daily on the Internet (this problem is identical to other types of innovation such as Uber or Airbnb). It will be necessary to innovate in terms of methods.

In the future, we will simplify tests and modify their monitoring. Different experimental plans could be used (e.g. randomizing doctors or periods rather than patients).

Attempts to integrate clinical trials into routine care (point-of-care trials) will also be made. It will involve "embedding" the trials in registries (registry-based RCTs) or cohorts (cohort-based RCTs) and developing "direct-to-participant trials".

11.1.7. *Exploitation of existing records*

Attempts will also be made to use data that is already collected for other purposes to measure key impacts (such as routinely collected clinical or administrative data, for example) rather than launching an original and expensive collection of specific variables. With e-epidemiology, the incremental cost of including additional patients into a cohort or a trial can come close to zero.

This is not possible for all treatments, but applies to interventions that replicate current clinical practices: treatment strategies (including surgery), drugs in clinical practices, drugs for new indications, interventions on CE marked medical devices and those already in use.

Examples are provided from the literature:

In case of an infarction, only inpatients are followed. This only represents 70% of affected patients. We followed 6,000 patients over 2 years, around 150 variables, including doctors, the hospital and the real world. This monitoring and prevention allowed the cost to be lowered to 50 dollars per patient. It is only necessary to envisage in the approach a place to randomize.

In addition, we must go from site-based trials to direct-to-participant trials. We will also conduct site-based trials, in multiple sites, but with a single coordinating center, which remotely assumes all its functions. A preliminary experience was conducted in California, which was cheaper and less disruptive (for professionals). Direct recruitment is 30–60% less expensive for large studies and less disruptive for patients, corresponding to "real life".

Cohorts – or "e-cohorts" – are developed to make this possible, which allows for the pooling of research costs. People are also asked to bring their own communicating device/object which will be used to collect data (which also reduces costs).

Many studies will be part of collaborative research, around patients, the environment and administrative databases, with routine data collected during the care process.

11.1.8. *Example*

This is the case of the ComPaRe[4] cohort, which concerns patients with chronic diseases. It is an e-cohort of 200,000 patients, which can be followed over time. The initial target period was 10 years and we can extract specific cohorts. Such a cohort of chronically ill patients makes it possible to develop knowledge on both multi-morbidity (association of several chronic diseases) or on each particular chronic pathology, etc.

11.1.9. *Conclusion*

We are on the cusp of developing new methods. The rules will necessarily be different from those of the drug industry, with reduced costs; the level of evidence to be achieved should be variable depending on the stakes and risks.

11.2. The key elements to behavioral change: from simple tools to complex intervention support – the example of prevention

11.2.1. *E-health and prevention: which evaluation for which intervention?*

We are at a stage of booming development for mobile applications and connected objects (SDApps) in healthcare. The O.A. specifically focuses on the conditions of effectiveness in their application during primary prevention. These tools are intended to favor the modification of individuals' behaviors according to their own characteristics, their environment (including their socio-economic conditions), as well as monitoring their health status, to induce behaviors more adapted to their health.

The challenge of research work, in which EHESP is engaged, involves modeling a theory of intervention. It involves analyzing connected objects and applications in a qualitative manner, and being a source of recommendation for application creators (as well as for users). Currently, users are involved in the work, including on an international scale. It is therefore essential that information about analysis is given to professionals as well as to the public; there are grids for doing this that are primarily

4 Now called "Nous Patients" (US Patients): https://nouspatients.aphp.fr/.

technical. However, there is a lack of theoretical models of intervention, where the users, within the framework of educational approaches, construct their position and decide, potentially in collaboration with a professional, the modalities of actions that concern them.

We highlighted in the introduction the importance of digital integration in society in general and in the medical field in particular. However, it is important to remain vigilant concerning the impact of these tools on populations and the type of individual who can benefit from them. Those who need them the most will have the least access to them, even if it is a simple object that can be put into their pocket.

11.2.1.1. *What is the real added value of these e-tools?*

Some of the questions that should be asked are the following:

How do they contribute to changing behavior according to the environment and the living conditions of individuals? It is not because a phone can count footsteps that individuals will walk more. If it were so simple, nobody would start smoking and everyone would do the amount of exercise required to remain healthy. If tools were a solution to the question of behavioral change, that would be known. Another question is then:

What place is there for individuals according to their psycho- and socio-environmental conditions?

The practices described in the literature refer to the following aspects and we will cover these again at a later point:

– these tools enable us to perform the self-measurement of health. They are socialization tools, a means of self-validation, by sharing collected data;

– there are three modes of use (surveillance, systemization or regularity and performance);

– there are new avenues for evaluations opened by this type of tool;

– can we envisage a new model of e-prevention? In reality, the current models are very poor.

11.2.1.2. *Self-measurement of health*

Technology allows an objectification of one's health and behaviors (the "quantified self") and especially:

– the monitoring of one's health and/or behaviors to quantify one's health activities or constants;

– the development of self-knowledge thanks to sensors, measurement objects (TA, pedometer, etc.).

One of the problems with medical data is that while an isolated measurement does not provide a piece of medical data (weight, size, etc.), as soon as these results are crossed, they become medical data.

In terms of prevention, technology enables the collection, measurement and comparison of the variations in biological, physical, behavioral and environmental parameters concerning everyday activities (sleeping, eating, exercising, etc.). Technology can:

– improve well-being and maintain or improve consumption-type health practices (smoking, drinking alcohol, balanced diets, etc.) or activity (work, leisure, exercise, etc.);

– record and analyze data (biomedical: TA, heart rate, etc.) in the context of a doctor–patient relationship concerning a specific risk.

It is important to distinguish data which is used by the individual, and data which serves as a support for exchange with a third party. In this case, it is proven that the intervention of the individual, via their smartphone for example, leads to an increase in efficiency. Sharing information with a third party strengthens the relationship of trust and accredits the value of advice given.

According to [ARR 13] and [PHA 13], the permanent self-measurement of one's health causes a perception of the modeled body in an essential technical report determined by quantitative data. We tend to conform to the norm. We may be ready or not to enter into such a logic.

Self-measurement highlights a behavior objectified by a self-constructed individual strategy (I can commit myself because of this measurement in new physical activities)… but this tends to never last long [MON 12]. The exercise bike often ends up as a coat rack. We change at the beginning, but at the price of big efforts. The levers of behavior are more complicated. If we do not know things, there will be no change in behavior. However, if we do dispose of such knowledge, this does not mean the battle is won.

Finally, self-measurements modify the boundaries between the fields of well-being, health and care: we tend towards a continuum between the normal and the pathological. Each individual develops their own ideas. The intervention of a professional tends to weigh *things down*.

11.2.1.3. *The connected object as a tool for socialization and a means for valuation (Martin, 2014)*

It is possible to share one's collected and analyzed data, and to belong to a connected community of users to value one's efforts and develop a sense of comfort in different ways. The user of the tool can be influenced by the behavior of the group. They can become an element of a social group [MAN 93] and the behavioral dynamic is linked to the dynamics created by the group.

11.2.1.4. *Three usage modalities*

Surveillance

The measurement of a risk and the notion of threshold have central roles, with the need to distinguish between snapshots and averages. Supervision is necessary: the risk is usually defined externally, often by medical standards (e.g. BMI). This modality is not centered on actions but on self-monitoring. The results can sometimes be a source of anxiety that does not promote sharing (bad results are my fault, I do not want them to be known…). The advice exchanged in the connected groups must answer to a logic of supportive mutual aid.

Systematization or regularity

It is about replacing a bad habit by a more health-promoting behavior (e.g. quitting smoking and adopting a new tobacco-free lifestyle and dietary measures). The central element is the regularity that comes from the motivation for an action or change. The exchanges in connected groups are designed to promote encouragement, but some prefer to avoid dealing with peoples' different opinions. For example, in physical activity, if the user is not at a certain level, they may get discouraged and give up. There would be a positive reinforcement for the top performers, whereas those who get bad results will disconnect.

Performance

Different measurements help to determine one's own goals, improve motivation and performance. However, this is not always the case. If it takes 10,000 steps to unlock the "badge", it is not stimulating: it would be more logical to set it to 3,000 steps and be sure to achieve that goal. It is then necessary to be able to adapt the tools. The connected object is only one tool. Exchanges of experiences in the connected groups promote competition but can also scare people off. Conversely, the challenge can influence the norm.

In practice

A "standardization" of individual activities does not exist:

– exchanges between users are rare;

– the alignment of practices between users is not a real expectation of them;

– measurement practices tend to decline with time [PHA 13]: ⅓ stop before 6 months, 39% of commercial applications are used less than 10 times.

Technological media coverage and social mediation [ARR 13] have made it possible to renew forms to highlight and describe ourselves personally. This is an opportunity to communicate according to new codes [AGU 09, CAL 14].

11.2.1.5. *New avenues for evaluation*

There is an effectiveness regarding these tools: studies have revealed key functions for effectiveness: fast, reactive and intuitive use tools are necessary; an interest and effectiveness must be perceived by the individual; the user must trust the content; a human presence increases effectiveness but only if the individual is recognized as legitimate (expertise, trust, listening, taking time, etc.).

Tools that only conduct measurements have a limited effect; the Internet is often used to reduce costs, but it is more effective when it is combined with coaching over the phone.

An ethical question persists: what health inequalities exist due to the use of these tools (few studies explore this question)? Connected tools can increase health inequalities according to at least three points:

– financial access to technology [ENG 98];

– technological limitations to enable efficient access (territorial inequalities) [VIS 07];

– the individual characteristics that influence the access and use of connected tools (culture, education, values, etc.) [KRE 04, BER 01].

The lacking presence of behavioral change theories: a main focus of evaluation: only 20% of publications cite theories of behavioral change and none are detailed in their implementation or evaluation. Sometimes the taxonomy of change techniques is cited [MIC 13] but no details are provided on the implementation of these techniques. When they are cited, five main theories appear: goals and planning, monitoring, shaping, knowledge and social support.

Nothing is said about the modalities that facilitate change: the main goals are to educate and promote health.

Main lesson: go from the role of actor to user: the most effective tools are those that allow an individual to find ways to change their behavior in their environment: allow them to identify the individual and environmental factors that influence their health and enable them to act upon them (which are different from self-quantification and information exchange).

11.2.1.6. *Discussion*

The central focus of this chapter remains the understanding of transformation mechanisms of individual behaviors. In fact, it remains unexplored, especially certain interesting areas concerning connected health. Focused on prevention, it does not touch on patient follow-up. Similarly, the question of the individual's relationship with doctors is not addressed as such, nor are the relationships between professionals. Reflections have not yet been initiated at this level.

The literature sometimes evokes social networks, especially in terms of prevention. Indeed, applications and connected objects are often associated with such networks. Professionals mobilize them among others to promote health and promote prevention tools.

An implicit hypothesis is that the individual has a relationship of trust with the tool. The confidence that the individual has with the professional with whom they communicate is necessary, but it is also necessary to trust the measurement provided by the tool. This was demonstrated and it shows that these elements mutually reinforce each other.

There needs to be a level of trust in the content: this is a requirement for this to work. But do we give the actors the means to trust? The builders do not share the studies they lead. This is one of the problems of the international network for the evaluation of connected objects. It is necessary to recover data, even the most basic – such as the number of footsteps – to build networks and accumulate knowledge. Living Labs have a role to play in their ability to mobilize user groups for validation studies with user data taken from the design stage.

The research presented does not address ergonomics, but this subject is discussed several times elsewhere in this book, especially in section 7.1.1[5]. It is indeed important to be able to use a connected object in a very simple manner. If a connected voltage application requires each user to turn it on and enter information, etc., they will not do it 10 times a day. It is the same for a connected scale: everyone simply wants to get on it and see that it works!

5 Regarding ergonomics, see also [PIC 17a].

12

Interdisciplinary Perspectives

Connected healthcare is largely the integration of knowledge, technologies and practices developed elsewhere in the field of healthcare. Analyses of this phenomenon are often limited to the visible part of this movement, in other words to the technologies themselves. For example, "connected objects in healthcare" are the subject of many reflections. Indeed, the regulatory framework of this sector, which guarantees essential requirements of citizens and their privacy, must find new forms, and technologies must adapt. However, other disciplines such as design, management sciences, innovation management and engineering also find opportunities to transform themselves, to adapt, develop, and thus become catalysts for a transformation in the healthcare sector which is needed by public health and the economic sector of healthcare products.

12.1. Anticipating connected objects in healthcare: combining patient–user leaders and design

Design projects in healthcare and autonomy are most often based on an individual "conventional" research project where the professional proposes design solutions to a problem that they have identified.

Design approaches co-constructed with users and with all actors and stakeholders are more relevant in the context of healthcare and autonomy. These approaches experiment with co-designing methods, implement iterative and interdisciplinary processes, and integrate contributions of

---

Chapter written by Mathias BEJEAN, Gaël GUILLOUX, Robert PICARD and Hervé PINGAUD.

knowledge specific to healthcare and concerning autonomy according to the subjects discussed.

Design professionals learn how to determine the relevant places, times and modes of intervention by placing the patient at the heart of their arrangement. They refine their designer know-how, broadening their knowledge in a particularly sensitive and complex domain.

These approaches also have the advantage of raising awareness among designers of a domain in which skills are organized into specialties, which does not always promote a general consideration of the issues experienced by the user and their entourage. Design then opens up new perspectives. It consists not only of offering solutions but also about being proactive in putting healthcare and autonomy into a framework of societal well-being.

This leads to a new approach to design in healthcare and autonomy. It depends on the points of entry which are critical in ensuring its success:

– the question of the level of knowledge needed to conduct a design project in healthcare and that of its transmission to designers. The translation of this knowledge into tools and methods usable by designers is one of the issues concerning the articulation of knowledge, both scientific and that specific to design, to determine the attachment points for a design approach;

– the interest in creating a map of actors to locate solutions (roles, missions and interrelations around the intervention arrangement opening up real innovation);

– a process aware of upstream development which goes from immersion to co-design of the solution involving all stakeholders (based on the link between methods and tools used in design and anthropology).

On this basis, it is necessary to pursue research in the development of tools and methods specific to healthcare design (which remain elementary and require scaling), according to the designer's specificity of knowledge and position in this domain. The in-depth study of a co-design approach seems particularly appropriate in a sensitive context where the perception of technical competence is ambivalent. Technical expertise is expected by the patient, but the technical universe in which they find themselves is also an area of insecurity. The methodological implication for the design of healthcare is to engage processes facilitating the mediation in a given context (personal history with the system) and to convey an appeased perception of complexity by using appropriate arrangements.

The approach must be structured by integrating all the public and private actors of healthcare and autonomy within the considered territory, which fit the situation and are within the studies' context.

More specifically, if we are interested in connected objects and interactive design (tangible interfaces, data visualization, virtual reality, etc.), understanding the current context is important.

Supported individuals show little enthusiasm for the use of connected objects (scales, blood pressure monitors). As a result, companies have been "beating around the bush for years without seeing the end of it" (Philippe Metzenthin[1]). Metzenthin recalls that the burden of French regulatory framework is significant[2], and private doctors' lack of willingness to get involved with its use or support the deployment of medical devices ("I do not have the time", "not paid for it") are two contextual obstacles which are often cited and do not favor the diffusion of connected healthcare equipment, especially in the patient's home. In France, supporting healthcare through technology is directly dependent on the investment made to reduce operating costs and improve the quality of service. Owing to this, few companies in the sector can and want to engage in this type of development.

First, there is a "judicialization" problem concerning the relationship between the patient and the healthcare professionals. Professional insurers for doctors refuse to ensure diagnoses based on these objects (not codified, not certified, etc.). French practitioners are afraid of misdiagnoses whose consequences would not be covered by insurance companies, which are amplified by laws surrounding class-actions[3]. In addition, a perceived "paperwork phobia" is linked to the digital world, which, on the one hand, disrupts habits, but, on the other hand, may expose the practitioner as being the only one, in a network of carers surrounding their patient, to use the system. Furthermore, professionals can use various or even different

1 Interview with Philippe Metzenthin, administrator at the Française de Domotique, February 2018.

2 The European framework, which imposes itself on France, is not less so.

3 The class action: flagship measure of Hamon's law on consumption passed in February 2014 by the French parliament, entry procedures, 1 October 2014, the group action allows consumers to organize themselves when faced by the same prejudice. Class action proceedings in the USA and UK have a far longer history.

solutions, environments and technologies. This does not simplify the digital relationship regarding a global chain of actors working collaboratively.

Three entry points regarding the proposition of connected objects in healthcare seem to favor their future diffusion:

– refundable telemedicine;

– motivated healthcare teams involved in healthcare innovation;

– avant-garde patient communities: "patient-users who are leaders of their pathology".

It remains necessary for reimbursable telemedicine and tele-surgery acts to be better understood by insurance companies. They are still perceived as rather vague and insufficiently defined. The possibilities of being paid are limited and supervised, in the context of healthcare professionals. In the short term, the context is not facilitative.

Some care teams and hospitals are motivated to implement them around specific care or particular pathologies (for example, the connected stethoscope). However, the advantages of these objects are poorly validated, especially since non-connected tools, which are just as effective, can easily be implemented. The experiments are few and upscaling in the short or long term is questionable.

However, certain patient-users exert significant pressure, linked with their pathologies, such as HIV, diabetes, cancer or SEP, to "force" the use of more modern and technological means. They are expert patients who appropriate these new technological tools which make their lives easier. They are the first promoters of these technologies and tools: active communities, communicating communities, fond of sharing experiences, etc. They are a type of community that has taken control because they have mastered the associated medical processes. For example, home dialysis patients go through a home-appropriation phase and know how to conduct first-level maintenance.

This last context of trust, which seems to possess a natural thirst for integration or even adhesion, seems to us, in the very short term, to be very fertile ground to co-construct solutions that will concretely meet needs, and to disseminate technical equipment and connected technologies from peer-to-peer, between patient–users. The in-depth study of a co-design approach seems to be particularly appropriate for these patient–users, to then transfer it to other usage contexts and healthcare practices.

It is all the more true as the adoption of a system remains strongly influenced by climatic and geographical conditions, the level of regulation, and the culture of familial, social and participative life (more developed in the Global North) in the environment. The integrative capacity of the design approach, considering the limits and opportunities of a situation or context to reach a materialized conclusion (solution), is therefore a real asset.

12.2. Maturing new connected healthcare solution concepts: towards a controlled integrative approach

The purpose of this section is to present the challenges of an "integrative" approach to the maturation of concepts in the field of connected healthcare. The word "integrative" means an approach considering the multiplicity of an innovation situation without belittling it; therefore, the organization of stakeholders' design and imagination efforts does not reduce the multiplicity of their own knowledge, experiences and temporalities. After having recalled the strategic nature of concept maturation activities, the chapter draws on a case from another sector, the space sector, to draw lines of work in the field of connected healthcare.

12.2.1. *Strategic concept maturation activities*

In the field of innovation management, streams of "radical innovation" [OCO 08, OCO 06, SLA 14], "open innovation" and "collaborative innovation" [CHE 03, LAU 06, WES 14] have paid attention to new organizational forms of innovation, their strategies, their processes and their business models. In particular, an important aspect of recent literature concerns the more "upstream" phases of exploration, including the formulation and maturation of innovative concepts [MAR 17, NIC 15].

During these maturation activities, which no longer consist of only evaluating and selecting ideas, as was advocated by "new product development" inspired by project management [COO 94], but also structuring complete management systems [OCO 08], intended for "mature" concepts and the organizations which sustain them [GAS 10, HOO 16], to transform the starting intention into a value proposition implemented into a new ecosystem of uses.

In this context, the strategic challenges are that of formalization and instrumentation of new "upstream" processes which are more agile and participative, integrating a wide range of stakeholders, both internal and external, including the end-users. Owing to this, many contemporary approaches to the strategic management of innovation mobilize design in order to generate new meanings and experiences for innovative concepts, products or services far upstream in their development [PIN 17, VER 08, VER 11], thus crossing into the field of design studies.

However, even as forms of design and innovation are increasingly open and participatory in practice, many difficulties remain in formalizing and organizing the process of collaboration between heterogeneous actors in a situation of innovation, particularly in more upstream phases. Among these difficulties, one is to articulate the "clock" of innovation projects, with the time of their "maturation", that is to understand, and not only to explain, the qualitative enrichment of complex, multifaceted and innovative concepts.

12.2.2. *The enrichment of concepts: "clock" time versus "maturation" time*

As Béjean and Drai [BÉJ 18] remind us:

"In management, the notion of time has long been restricted to a 'clock-like' definition, meaning time is reduced to a spatial displacement along a straight axis. Project management tools, like Gantt charts, mobilize this quantitative and linear representation of time. In recent years, the vision of time has been enriched, particularly in the management of innovation activities. In the 'Labs', time is more akin to that of a concept's maturation; it refers to a second facet of time, as a change in quality, that the philosopher Henri Bergson named 'duration', i.e.: 'a growth from within, the uninterrupted prolongation of the past in a present encroaching on the future' (Bergson, 1938)" [BÉJ 18, p. 193].

Compared to the creation of innovative concepts, the time of duration, of *maturation*, makes it possible to grasp an evolution without juxtaposing moments ordered linearly in advance. It thus allows us to consider the qualitative enrichment of the object of collective enquiry or the effort of imagination and design. From a management point of view, it therefore seems difficult to only consider clock time, at the risk, otherwise, of locking up maturation activities into procedural "fossils" blind to the qualitative enrichment of concepts in a situation.

However, caution must be exercised in how to proceed because "experienced" time is not homogeneous for all actors in a situation of innovation. Seeking to reduce it to a single "collective actor" could reduce the multiplicity of one's own knowledge, experiences and temporalities; similarly, seeking to exhibit this multiplicity would risk revealing an illegible complexity to participants, as well as to the sponsors of the innovation process.

The problem of time therefore arises centrally in maturation activities. If the "clock" time tends to mask the various concept maturation "paths", ordering moments into a linear representation fixed in advance and reducing the multiplicity of internal exploration dynamics, it nevertheless allows the identification of "phases" in an innovation process (ideation, experimentation, validation, etc.) and thus to organize collective action. Is this an irreducible antagonism or is it possible to combine these two facets of time?

12.2.3. *An example: the space industry and the genesis of new mission concepts*

To advance the understanding of the difficulty stated previously, it is useful to use an example. This example does not aim to provide a definitive "solution" to our problem, but to guide our questioning in a certain direction. The use of this example must then be taken more as an invitation for reflection, rather than a call to replicate this case in other contexts. We will see elsewhere that such an immediate response would be contradictory with the aim of this type of organizational response to the problem of maturation.

The case presented is based on the experience of the Jet Propulsion Laboratory (JPL), a joint entity of the California Institute of Technology and NASA. For a more detailed presentation of this case, refer to [BÉJ 16], as well as to JPL publications available online [ZIE 13]. The interest of this case for healthcare is that design is also a long-term process. Furthermore, the space industry has had to deal with very challenging engineering problems, often involving very heterogeneous actors, including end-users.

Among the difficulties encountered by the JPL, we can thus provide as an example: a field of intervention beginning before classical architectural investigations can take place, the combination of "elements" of solutions from different suppliers with different technical experience, the maturation of complex and multifaceted concepts at the crossroads of the latest available scientific and technological knowledge, bringing the end-users and markets onboard very early, according to a structured and shared approach.

Here, we will be able to make the link with difficulties arising during the maturation of new concepts in connected healthcare: difficulties of upstream co-design, combining usage evaluations in Living Labs and autonomy with clinical investigation measures, exploration of complex polymorphic therapeutic solutions and so on. Finally, in the space industry as in healthcare, when the system is launched or disseminated in the population, we can hardly intervene: in healthcare, an intervention alters hypotheses and invalidates trials, all while creating risks of deteriorating clinical outcomes; in the space sector: "When the rocket takes off: we can no longer change anything!"

12.2.4. *Upstream phases: from sequential logic to concurrent engineering*

"Every space mission starts from a spark"[4]

In official JPL documents, we can read the following: a mission results from the coming together of three fields: the scientific field, which poses a question or expresses the need for measurement; technology, that is an invention or a new technology that may be more or less mature; and engineering, which develops a concept of spatial measurement. JPL intervenes at the crossroads of these three domains, as the "architect" seeking to integrate a "system of systems", as to not always be in one's comfort zone.

---

4 Kelley Case, JPL Innovation Foundry, Presentation, California Institute of Technology, October 2012.

In the 1990s, faced with difficulties in organizing new space mission projects, the JPL decided to reorganize its upstream activities to pass from a mainly sequential logic to a concurrent engineering logic, thus setting up a "concurrent engineering center" and creating "Team X" within the JPL, which would become known worldwide. It brings together 20 engineers organized by "key mission subsystems" ("Power", "Telecommunications", "Propulsion", etc.). A typical "upstream" study consumes 300 to 400 hours of engineers' time and develops 1 to 3 "point design missions".

However, despite this reorganization and many successful projects, difficulties persist for the most upstream phases in the formulation of project requirements ("early mission formulation").

To be effective, concurrent engineering requires a mature understanding of technical and scientific specifications. But for example, the JPL must often interact with clients (or technologists) whose thinking is not mature enough.

These problems result in quality, costs and delivery deviations for numerous study projects. In the early 1990s, Werner Gruhl, member of the NASA's Comptroller office, provided several economic studies showing that for a given mission, if 10% of the total time is not allocated to the initial formulation phase ("definition percent" according to Gruhl), excess costs for the mission will reach at least 30%. This study resulted from the analysis of more than 20 real missions and left its mark.

12.2.5. *Concept Maturity Levels (CMLs)*

At the JPL, new organizational challenges will continue to be identified. Consisting of formalizing a new mandate for activities of maturation and incubation, thus recognizing their strategic role; setting up new processes and dedicated teams; developing tools and methods to explore and select promising concepts while controlling the time of exploration and the resources involved; and having a common language to evaluate the maturity level when exploring a given concept.

These new challenges have led to the invention of new metrics, notably, Concept Maturity Levels (CMLs). These metrics are inspired by the Technology Readiness Level (TRL) previously developed by NASA in the 1980s, allowing for the evaluation of the robustness of knowledge for a given technology at a given moment. TRLs are a tool for project organization and communication, and have become essential today. Recognized as constituting a generic language, TRLs are used commonly in tenders, briefs, conferences, literature reviews, etc.

However, where TRLs tend to consider that only the quality of the demonstration (or proof) makes it possible to construct project robustness, CMLs introduce the innovative idea that the quality of exploration also allows robustness to be constructed, particularly in upstream phases. In other words, TRLs belonged to a logic of mastering convergence processes, while CMLs also integrate the command of divergent processes. In addition, CMLs integrate three "drivers" – need, technology and programmatic aspects (cost, strategy, organization, etc.) – while TRLs only considered technical and scientific aspects.

Thought of as a common generic language, CMLs therefore aim to allow the evaluation of a concept's maturity to make a hierarchized selection of missions to support. They thus cover the mission concept formulation phase, from a "cocktail napkin" (CML 1) to NASA PDR norm[5] (CML 8):

- CML1: Cocktail napkin;
- CML2: Initial feasibility;
- CML3: Trade space;
- CML4: Point design;
- CML5: Baseline concept;
- CML6: Integrated concept;
- CML7: Preliminary implementation baseline;
- CML8: Integrated baseline.

5 Preliminary Design Review: "The PDR demonstrates that the overall program preliminary design meets all requirements with acceptable risk and within the cost and schedule constraints and establishes the basis for proceeding with detailed design. It shows that the correct design options have been selected, interfaces have been identified, and verification methods have been described. Full baseline cost and schedules, as well as all risk assessment, management systems, and metrics are presented" (NPR 7120.5D, p. 30).

12.2.6. *An organizational response to the problem of "maturation"?*

Following the development of CMLs, a "diamond process" was set up at JPL to organize maturation phases CML 1 to CML 8, without having to start a study at CML 1 and/or end at CML 8. The diamond process has also included two new teams, the "Eureka Team" (CML 1) and the "Rapid Mission Architecturing Team" (CMLs 1 to 3), gathered in a new location, the "Innovation Foundry", all forming a structured organizational response to problems involving maturation activities.

However, if we clearly see the value of the JPL's organization response, could it be reproduced in another context and, if so, how? Is it possible to use the CMLs as they are? Additionally, could they create any other difficulties?

A response pathway comes from research [BÉJ 16] conducted in France in the context of the *Plateau d'Architecture des Systèmes Orbitaux* (PASO), a transverse organization in charge of preliminary studies and 0 phases within the *Centre national d'études spatiales* (National Center for Space Studies, CNES). At the time, PASO's role for the CNES was comparable to JPL's for NASA, although with significant differences in terms of size and status. Aiming to analyze the "formulation of requirements" for upstream phases (named "Phase 0" at the CNES), the research helped formalize CMLs for CNES. Several lessons were learned from it [BÉJ 16].

12.2.7. *The case of healthcare*

Among the lessons learned following the research conducted at PASO, it can be pointed out that CMLs cannot be perceived as universal metrics applicable to all fields, but that they require a development effort specific to the type of concepts studied. On the contrary, while there is a clear interest in "maturity metrics" to address the multiplicity and temporality of the maturation processes in upstream phases rather than a solution to the maturation problem, the CMLs are more of an invitation to commence questioning.

What about in healthcare? There is an observable great difficulty for healthcare authorities and public policymakers to address the disruptive innovation of connected objects. Therefore, for example, the expected funds for financing innovative medical device (MD) projects have only resulted in a very minute number of projects. According to J.-Y. Paille[6]:

"The innovation package, which offers financial support for innovative treatments in healthcare excluding drugs, is not often utilized [...] Companies denounce expensive and complicated procedures that are not suitable for medical device manufacturers".

However, supporting innovation to compensate a stall in development of new medicines is fundamental, both in terms of public healthcare to continue improving population health and in economic terms, for companies to develop in this sector.

An interpretation of this difficulty is that the upstream phases of therapeutic solutions integrating digital technologies are insufficiently understood and poorly understood. Traditional clinical trials are failing (see section 11.1) but no alternative frames of reference exist for responsible decision making.

In the field of connected health, CMLs should therefore provide a different approach to the upstream co-design difficulties, the way real-life data is articulated with the results of clinical investigations, or even explore complex polymorphic therapeutic solutions at different levels of maturity, avoiding for example the "techno push" phenomenon, as applicable to the spatial domain and to connected health.

Throughout this book, it has been shown that user, patient and professional engagement, or even a citizen concerned by their heath and the anticipation of their ability and willingness to delve into these new approaches are major success factors. The CML approach to health must address this concern. It fully justifies the engagement of the Forum LLSA in this work, with the joint mobilization of Living Labs, patient associations,

---

6 "Santé: le forfait innovation a du mal à décoller", *La Tribune*, 12 July, 2017. Available at https://www.latribune.fr/entreprises-finance/industrie/chimie-pharmacie/sante-le-forfait-innovation-a-du-mal-a-decoller-743597.html.

and clinical investigation in technological solution centers, alongside manufacturers and integrators.

12.3. Observation of the transformations in engineering by a Living Lab: some elements of reflection

12.3.1. *Evolutions in engineering of a Living Lab anchored in its territory*

There are no Living Labs in reality without considering their territorial roots. The Connected Health Lab (CHL) came about in Castres Mazamet cities, in 2015 and affirmed its roots. Effectively, CHL has the specific feature of arising in an ecosystem of higher education and research, within an engineering school specializing in e-health, a vector of economic development of this native country. After three years of experience in this context, we offer a first round of feedback and deliver some observations about the new roles played by this Living Lab in the activities of the academic structure that incubates it.

The CHL is a system that delivers services at the interface between the internal actors of the school and their industrial, academic and institutional partners. In this position, it carries out missions where the practices and engineering approaches are included in the service delivered overall. The CHL is also a platform that recreates an environment that is close to the realities of hospital and home care. Its purpose is the progress of connected health on the patient pathway. When engineering students, PhD students, graduate students and researchers participate in CHL activities and co-construct simulations where technology is assessed in life-like conditions, these are pragmatic and intra-murous experiments that are realized and contribute to progress in terms of individual and collective learning. The Living Lab is then a special location for thinking about engineering practices in the target application sector; its stability allows for long research periods and therefore allows us to draw assumptions about future changes in engineering practices. It must be noted that this observation is slightly biased since the use of CHL influences the course of events and such an intrusive character probably influences the true course of practices.

12.3.2. *Engineering professions: what are we talking about?*

12.3.2.1. *Engineering: a definition*

A fairly common definition of "engineering", in other words the engineering that is still perceived in everyday life by society is as follows:

DEFINITION 12.1.– Engineering covers all range of functions, from design and sizing of things to building or manufacturing process layout, as well as equipment control of technical objects or industrial plants.

12.3.2.2. *Knowledge and missions of the engineer*

In France, the *Commission des Titres d'Ingénieurs* (CTI), the engineering qualifications committee, has as its mission the evaluation of higher education institutions for the engineering professions. It also continuously maintains a generic profile of what an engineer should be, i.e. involved in a trade which complements the needs of our society. For CTI, engineers ensure within organizations, mainly companies, a wide range of functions in the following categories:

1) fundamental and applied research;

2) studies and design, consultation and expertise;

3) production, exploitation, maintenance, trials, quality, security;

4) information systems;

5) project management;

6) customer relationships (marketing, trade, after-sales services);

7) management;

8) education and training.

The relationships that exist between science and engineering, two worlds that are all too often contradictory in French higher education, have been thoroughly debated. The first attempt of scientific classification in fundamental and applied sciences proposed by Auguste Comte was then developed by Herbert Spencer into four classes: mathematics, physics, biology and natural sciences, and finally moral sciences (also known as humanities). The order of presentation of the first three classes progressively shifts from the abstract to the concrete. In its operational reality, engineering is initially broken down into specialties often rejected in relation to these scientific disciplines. You must deal with mechanical engineering, electrical engineering, process engineering, agronomic engineering, etc.

12.3.3. *The challenges of engineering professions*

Because companies produce more and more complex objects and emphasize, for reasons of competitiveness, the relationship between commercial products and associated services, a new wave of engineering profiles has emerged featuring an affirmed multidisciplinary vocation. Disciplines such as genetic engineering, bio-mechanical engineering, biomedical engineering, industrial engineering and information systems engineering, for example, have gained interest and seem to be well established nowadays, as they appear attractive to students searching for high-tech jobs. These designations go far beyond the scope of a decline in fundamental sciences and promote an integration of multiple skills for a more precise and sharper, specific application field, often approaching more spontaneous scientific research and innovation.

12.3.3.1. *Understanding the uses*

Bernard Charlès, Director of Dassault Systèmes recently expressed his opinion in an article in the French newspaper Le Monde:

"Value is no longer the products, but its use: we are in an economy of experience, where subject and object are held together. The industry of the 21st Century surpasses the flow of parts to the benefit of flows of uses and virtual models, in a data economy [CHA 18]".

This is a key point, the time when the disciplinary corpus of engineering training constituted its main footprint, and the hard sciences its heart of heritage, may have reached the limits of its existence. The current era would be optimal for the generalist engineer (hyper specialization tends to appear only at the end of the learning process as refinement for the application field). Increasingly open to other knowledge, any engineering education is invited to focus on a new type of skill: learning to learn.

12.3.3.2. *Developing skills*

If scientific culture still remains a strong basic foundation of engineering knowledge, with time constraints, it is difficult to teach everything to a student in three years, however brilliant they are, with just over 2,000 hours of training time. New learning modalities should emerge. A profile must be developed in which the ability to work with others is considered of prime interest. The students must learn by participating in multicultural and diverse groups, and assert by the quality of their contribution jointly with other trades. And therefore, in any undergraduate training program, inductive pedagogy has to take precedence over deductive pedagogy. In this social dimension with increased collaborative work, the engineer must simultaneously master their scientific and technical production and the organization of group work through project management. Enterprise-based learning for students with apprenticeships and also for project-based learning for those who have student status (that is, non-apprentice students) have become the focal point towards which the lessons converge. From experience, it is the teacher who must reduce their theoretical courses, their directed and practical work to a fair minimum, and are encouraged to provide the basics, to give sources of knowledge in books and on the Web to which the students will refer, when they have not already found them on their own. Roles shift to a new form of control, not of knowledge, but of the ability of the student to acquire knowledge, to seek the right information with relevance, efficiency and lucidity. Would the target become more important than the path taken to reach it? Students must demonstrate their abilities through their project work. Therefore, to do it with a Living Lab within reach is an opportunity that everyone seizes.

12.3.3.3. *Training the healthcare engineers of tomorrow*

L'Ecole d'ingénieurs en Informatique et Systèmes d'Information pour la santé (French School of Engineering in Computer Science and Information Systems for Health) was designed at the crossroads of these paths, with a specialization in a personal service-based economy which was deemed as a niche at its initiation, and also with an awareness of the expectations of its young audiences and the expected difficulties that will arise if we decide to design a profile by just filling our brains with maximum levels of knowledge. Indeed, courses involve applied sciences with a fair proportion of knowledge about the human body, knowledge that evolves very rapidly. The fact remains that working directly on humans as technical objects and understanding them with a systemic point of view is revealed to be a source of concern for most CTI experts. How deep will you have to delve into knowledge in order to create? Its integration into the healthcare organization is another source of concern. How can we imagine the professional trajectories of graduates in a medical business world led by doctors whose appetite for information and communication technologies is still an essay topic? It is in the search for answers to these questions and the associated brainstorming that we must situate the birth of the idea to build the CHL, as a means of resolving divides by collaborations sought out with the medical world and with the idea of developing teaching practices in line with the complexity of topics covered.

12.3.4. *New approaches*

What requirements might we advance for modern engineering? The answer can be produced relative to what it was. We want modern engineering to be more agile, more open-minded and faster:

– more agile by the ability to work in uncertain situations and with know-how allowing a progressive design of work;

– more open-minded by the ability to work as a team, to accept consensus, to share conceptual diagrams, lexicons and objects with others, and to respect each other;

– faster by the energy made to honor appointments in regard to delivered project milestones, agreeing to deliver viable minimal products of which faults are recognized, spreading the workload with teammates and smoothing it out in terms of progressive and risky development.

In project management bodies of knowledge, the life cycle is currently introduced as the V life cycle, then after becoming aware of the concept of risk, through iterative cycles, where the V cycle is systematically repeated. System analysis and user requirements are at the beginning of each V, allowing us to begin the drafting of detailed specifications of solutions in which functional characteristics are the main drivers. The specification of what the solution will be is naturally based on the perception of needs as they have been formulated (not necessarily as they are expected to be). Iterative life puts emphasis on the question of a smart evolution of project content, based on the understanding of needs. Agile project management methods have addressed this problem. They advocate a new team organization where continuous relationships between a product's prescriber and the developers put the process of achieving the solution under control. Often, these organizations minimize any form of subordination between team members to encourage the empowerment of each member and develop a sense of common work. This product prescription acts regularly as a permanent communication cycle between the expectations of the sponsor and the effectiveness of the team of developers. There must be a relevant questioning and clever play between actors: what must be done? What can be done? What has already been done?

Centering understanding on the user-centric approach, where the user is included in the project team, improves very sensitive transmission activity. User interaction makes it possible to evaluate the properties of the product very early on, whether they are functional or not (see Figure 12.1). Coming from the world of design and ergonomics, methods such as design thinking or user experience are now recommended in this context. A benefit observed with this type of approach is being able to stop the developments while the user judges that their needs are sufficiently covered, avoiding spending time on unnecessary developments. There are evidently some conditions to be fulfilled to make the most of these steps. The choice of an effective users is one example.

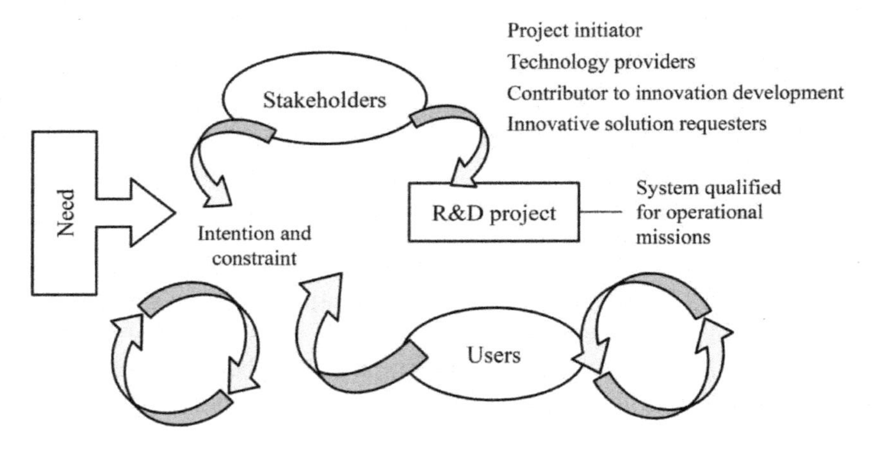

Figure 12.1. *Involvement of users in innovation, from intention to exploitation*

This evolution of project management practices feeds the transformation of the engineering profession. As Living Labs convey new ways to tackle both subjects and objects under study at the heart of their approach, their inclusion in the daily life of the engineer and their recognition by the profession are facilitated.

At CHL, all stakeholders take ownership of the places for different exercises, from acquiring English technical vocabulary as part of an international project, to creating videos capturing the context in which a project must change the course of things, as just two examples. It is in the same room that students will have to deal with different fragments of knowledge on a communication protocol, a database, a biomedical sensor and a semantic interoperability constraint. This access to knowledge is motivated by usage and by meeting in a motivating area. This motivation also extends outside, when we work in pairs with student nurses on ideas for developing new e-health devices. The student nurses in their reflexive practice must indicate the barriers they identify. The student engineer must understand them and propose an acceptable solution for them to be used in the future.

It is by direct interaction with users that the engineering process begins to unfold.

12.3.5. *Innovation sets new ways of acquiring knowledge*

Innovation is supported by an ongoing process that ranges from the perception of a source of progress to the confirmation of the progress achieved. This master process covers the management of the project portfolio, influences any strategic decision and is now included in organizations' balance sheets.

12.3.5.1. *Open innovation*

Open innovation forces the boundaries of the organization to allow the flow of bilateral knowledge. The challenge is to enrich the knowledge of others in order to promote internal progress. It can also be to support innovation by collaborating with other organizations that recognize the knowledge provided and reward it. Living Labs are actors of these open innovation processes that officiate the mediation position between use and forms of progress. It is a dimension that goes beyond the sphere of the project to enter a space where we must ponder upstream ideas, and where we must look at real-life uses downstream. Furthermore, this must be staged with cooperative rules of play where goals common to several partner organizations become drivers of development. The health sector is particularly fertile in this dynamic of open innovation. Clinical research has been involved in this area for a long time. The necessary evaluation of the effects on humans of all technical proposals, the critical nature of the target activity, the advice of the medical or medico-social body, the relationship to regulatory bodies and health insurance are among the factors that determine the degree of openness of innovation in health.

12.3.5.2. *Mastering a life cycle*

However, at the crossroads of innovation and knowledge, a question arises: is it possible to monitor innovation process with rationality? It is a big challenge to bring together all the knowledge to be conveyed during its running time. A conceptual representation of open innovation must be completed by the characterization of a life cycle, as was done in project management. If such a life cycle can only be defined at a high level of abstraction, it can be useful to trace all the acts performed in regard to an experience, to position the decision-making milestones and analyze the risks surrounding creative activities. We have launched a research project to start the design of this theoretical framework with Yosra Chaher's thesis subject (PhD thesis, Ecole Doctorale Systèmes, CNRS LGC, Université de Toulouse). It has produced a metamodel of open innovation and a first mapping of the life cycle that provides this sought-after rationality. Among the first results of this work, we now seek to identify good open innovation practices localized in the steps of the life-cycle. We also think that the framework is a wonderful tool for telling stories of open innovation in a comprehensive way. In this, we hope to progress in capitalizing on experiences and in fueling the work of case-based decision support tools.

12.3.5.3. *Living Labs: actors of change management*

Evidently, the launch of a process involves the selection and commitment of competent resources at each of its stages. The place and roles used by the Living Labs are subjects illuminated in a new light by the results of this research. By modeling system engineering principles, we can say that the relationship between the system to be made and the system enriches the rationalization of open innovation. The many practices and methods evoked here can be situated in an explicit epistemological process and can be mobilized wisely. It will then be up to Living Labs to play its part to enter into collaborative work with a recognized predisposition, where they will be solicited harmoniously within the framework of the new attributes of engineering, attributes that will become the markers of their modernity.

Conclusion: the Success of Conditions Linked to the Connected Health Approach

Citizen-connected health is a complex notion that we have sought to introduce to the reader.

In the first part, we have delved into the notion of citizenship associated with connected health. We have developed the aspects of responsibility and patient rights, the weak and/or frail, disabled or just individuals concerned about their health. These aspects are inseparable from the necessary evolutions on the side of the professionals and the organization of care. They are accompanied by new ethical considerations, which the *Etats généraux de la bioéthique,* in 2018[1] have highlighted in their new dimension of "techno-ethics".

The second part of the book mainly concerned the link between health and connection that has been delved deeper into. What can be collected about the individual or the immediate environment that can make sense for them, but also for the care team? How can this data contribute to the development of new knowledge, at the service of medicine and the search for new therapies?

Conclusion written by Robert PICARD.

1 A large public consultation organized in France to prepare the revision of the 2011 Law on bioethics.

We then looked at the ingredients of an efficient and sustainable citizen connected health: new approaches to design and evaluation of the solutions, the legal and regulatory frameworks and "soft law" and conditions of economic viability, not forgetting the technical requirements that are relevant to architectures, reliability and security – this goes without saying.

In the last section, we proposed some prospective reflections on the changes brought about by citizen connected health to clinical research and to prevention, and also, conversely to the contribution of new knowledge in innovation management and in engineering to the emergence of innovating solutions in this field.

This path has left alone more specialized issues, the difficulties of which we do not underestimate: new approaches in medicine made possible by the abundance of real-life data, conducting projects allowing the implementation of complex ambulatory solutions, to name just two of them. Specialized books exist on these subjects or will be published shortly. Our aim here has been to emphasize the importance of a cross-cutting vision of connected health, at the crossroads of many disciplines whose encounters and interactions have so far rarely occurred.

We hope to have been able to provide some elements of conviction on the importance of such multidisciplinarity. We are conscious of the efforts that this requires, both at the individual level, because of the scope of the field to be apprehended, and at the collective level, which requires the creation of a common language and, for each expert population, a willingness to interact with others while leaving one's comfort zone.

We would like, at this stage, to recall the favored framework of Living Labs in health and autonomy and the Forum LLSA that inspires France to take up this task. Indeed, each individual Living Lab is a network composed of a wide variety of stakeholders who are united by the desire to develop innovative solutions in collaboration with the public. As for the forum, it gives everyone the opportunity to widen the horizons of each individual Living Lab. Indeed, the Living Labs remain small teams (rarely more than 10 individuals). They focus on a territory, a methodological approach, a technique and a type of population. They do this by sharing experiences, problems, working on common elements and identifying patterns in these different spheres which are all vectors of progress.

The forum not only brings together the Living Labs, but also a certain number of other actors interested in this particular form of feedback on real-life practices in the field of health and autonomy. These actors are stakeholders of the forum's think tanks: current work is a reflection of this, which consists of industrialists, researchers and medical professionals.

The forum's recent status change, more precisely of the association which constitutes its legal structure, more clearly recognizes this proposition of value: the development and dissemination of new knowledge is now clearly part of its purpose; and its governance, which now relies on colleges that ensure that the diversity of members of this network is taken into account.

It remains to disseminate even further these observations, reflections and results. Beyond Living Labs in health and autonomy, beyond other members, already convinced by the extent to which they have joined, we carry the ambition to share these elements with the entirety of sector stakeholders, including the public decision-makers, in the various ministerial departments, agencies, commissions, specifically in healthcare, research and the industry.

May this book contribute to it.

References

[AAR 10] AARHUS R., BALLEGAARD S.A., "Negotiating boundaries: managing disease at home", *Proceedings of the SIGCHI Conference on Human Factors in Computing Systems (CHI '10)*, ACM, New York, pp. 1223–1232, doi: https://doi.org/10.1145/1753326.1753509, 2010.

[AGU 09] AGUITON C., CARDON D., CASTELAIN A. *et al.*, "Does showing off help to make friends? Experimenting a sociological game on self-exhibition and social networks", *Third International AAAI Conference on Weblogs and Social Media*, 2009.

[ALV 07] ALVAREZ J., Du jeu vidéo au serious game, approches culturelle, pragmatique et formelle, PhD thesis, Université de Toulouse, Toulouse, 2007.

[ARR 13] ARRUABARRENA B., QUETIER P., "Des rituels de l'automesure numérique à la fabrique autopoïétique de soi", *Les Cahiers Du Numérique*, vol. 9, pp. 41–62, doi: 10.3166/lcn.9.3-4.41-62, 2013.

[BAI 16] BAILLY S., DESTORS M., GRILLET Y. *et al.*, "Obstructive sleep apnea: a cluster analysis at time of diagnosis", *PLoS One*, vol. 11, p. e0157318, 2016.

[BAR 17] BARDRAM J.E., FROST M.M., "Double-loop health technology: enabling socio-technical design of personal health technology in clinical practice", *Designing Healthcare That Works*, pp. 167–186, 2017.

[BÉJ 16] BÉJEAN M., "Faire enquête", *Chemins et temporalités de l'innovation : les processus d'exploration de la valeur*, no. 2, Publications du CNES, 2016.

[BÉJ 18] BÉJEAN M., DRAI L., "Innovation, collaboration et droit", *Revue française de gestion*, vol. 43, no. 269, pp. 183–198, 2018.

[BER 01] BERLAND G.K. *et al.*, "Health information on the internet: accessibility, quality, and readability in English and Spanish", *Journal of the American Medical Association*, vol. 285, p. 2612, doi: 10.1001/jama.285.20.2612, 2001.

[BEY 17] BEYALA L., *Connected Objects in Health: Risks, Uses and Perspectives*, ISTE Press, London and Elsevier, Oxford, 2017.

[BON 13] BONNECHÈRE B., JANSEN B., OMELINA L. *et al.*, "Use of serious gaming to increase motivation of cerebral palsy children during rehabilitation", *European Journal of Paediatric Neurology*, vol. 17, no. S1, p. S12, 2013.

[CAL 14] CALDWELL T., "The quantified self: a threat to enterprise security?", *Computer Fraud & Security*, vol. 11, pp. 16–20, 2014.

[CHA 18] CHARLÈS B., "Intelligence artificielle : « Le concept d'industrie 4.0 propose de numériser le passé quand il faut imaginer l'industrie de demain »", Le Monde, 4 April, available at: https://www.lemonde.fr/idees/article/2018/04/04/intelligence-artificielle-le-concept-d-industrie-4-0-propose-de-numeriser-le-passe-quand-il-faut-imaginer-l-industrie-de-demain_5280588_3232.html, 2018.

[CHE 03] CHESBROUGH H.W., *Open Innovation: the New Imperative for Creating and Profiting from Technology*, Harvard Business School Press, 2003.

[CNO 15] CNOM, La santé connectée, enjeux et perspectives, Report, February 2015.

[CNO 18] CNOM, Médecins et patients dans le monde des data, des algorithmes et de l'intelligence artificielle, Report, January 2018.

[CON 05] CONSEIL D'ÉTAT ET LA JURIDICTION ADMINISTRATIVE, Sécurité juridique et complexité du droit – rapport public 2006, available at: http://www.conseil-etat.fr/Decisions-Avis-Publications/Etudes-Publications/Rapports-Etudes/Securite-juridique-et-complexite-du-droit-Rapport-public-2006, 2005.

[CON 13] CONSEIL D'ETAT ET LA JURIDICTION ADMINISTRATIVE, Etude annuelle 2013 : le droit souple, available at: http://www.conseil-etat.fr/Decisions-Avis-Publications/Etudes-Publications/Rapports-Etudes/Etude-annuelle-2013-Le-droit-souple, 2013.

[COO 94] COOPER R.G., "Third-generation new product processes", *Journal of Product Innovation Management*, vol. 11, no. 1, pp. 3–14, 1994.

[DEC 17] DE CHOUDHURY M., KUMAR M., WEBER I., "Computational approaches toward integrating quantified self sensing and social media", *CSCW: Proceedings of the Conference on Computer-supported Cooperative Work Conference on Computer-supported Cooperative Work*, pp. 1334–1349, doi: 10.1145/2998181.2998219, 2017.

[DHI 16] DHILLON J.S. *et al.*, "Designing and evaluating a patient-centred health management system for seniors", *Journal of Telemedicine and Telecare*, vol. 22, no. 2, pp. 96–104, 2016.

[DUN 13] DUNTON G.F., "Using smartphones for EMA", *E-Tools for Social Epidemiology Symposium*, Paris, 2013.

[ECK 13] ECKERT D.J., WHITE D.P., JORDAN A.S. *et al.*, "Defining phenotypic causes of obstructive sleep apnea: identification of novel therapeutic targets", *American Journal of Respiratory and Critical Care Medicine*, vol. 188, pp. 996–1004, 2013.

[ENG 98] ENG T.R. *et al.*, "Access to health information and support: a public highway or a private road?", *Journal of the American Medical Association*, vol. 280, no. 15, pp. 1371–1375, 1998.

[GAS 10] GASSMANN O., ENKEL E., CHESBROUGH H.W., "The future of open innovation", *R&D Management*, vol. 40, no. 3, pp. 213–221, 2010.

[GRO 14] GROSS O., GAGNAYRE R., "What expert patients report that they do in the French healthcare system, and the competencies and personality traits required", *Therapeutic patient education*, EDP Sciences, vol. 6, no. 2, 2014.

[HCA 12] HCAAM, "Avenir de l'assurance maladie : les options du HCAAM", available at: http://www.securite-sociale.fr/IMG/pdf/l_avenir_de_l_assurance_maladie_les_options_du_hcaam.pdf, March 22, 2012.

[HCA 16] HCAAM, Innovations et système de santé, Report, vol. 2, 2016.

[HOO 16] HOOGE S., BÉJEAN M., ARNOUX F., "Organising for radical innovation: the benefits of the interplay between cognitive and organisational processes in KCP workshops", *International Journal of Innovation Management*, vol. 20, no. 4, p. 1640004, 2016.

[HUY 16] HUYGENS M.W. *et al.*, "Expectations and needs of patients with a chronic disease toward self-management and e-Health for self-management purposes", *BMC Health Services Research*, vol. 16, p. 232, 2016.

[HYN 09] HYNES D., RICHARDSON H., *What Use is Domestication Theory to Information Systems Research?*, IGI Global, UK, 2009.

[KHA 12] KHALFA N., BERTRAND P.R., BOUDET G. *et al.*, "Heart rate regulation processed through wavelet analysis and change detection: some case studies", *Acta Biotheoretica*, vol. 60, nos 1–2, pp. 109–129, 2012.

[LAU 06] LAURSEN K., SALTER A., "Open for innovation: the role of openness in explaining innovation performance among UK manufacturing firms", *Strategic Management Journal*, vol. 27, no. 2, pp. 131–150, 2006.

[LEG 11] LE GOFF-PRONOST M., PICARD R., "Need for ICTs assessment in health sector: a multidimensional framework", *Communications and Strategies*, no. 83, pp. 87–108, 2011.

[MAN 93] MANSKI C.F., "Identification of endogenous social effects: the reflection problem", *The Review of Economic Studies*, vol. 60, no. 3, pp. 531–542, 1993.

[MAR 17] MARKOVITCH D.G., O'CONNOR G.C., HARPER P.J., "Beyond invention: the additive impact of incubation capabilities to firm value", *R&D Management*, vol. 47, no. 3, pp. 352–367, 2017.

[MCK 15] MCKINSEY&COMPANY, "The Internet of Things: Mapping the Value Beyond the Hype", available at: https://www.mckinsey.com/~/media/McKinsey/Business%20Functions/McKinsey%20Digital/Our%20Insights/The%20Internet%20of%20Things%20The%20value%20of%20digitizing%20the%20physical%20world/The-Internet-of-things-Mapping-the-value-beyond-the-hype.ashx, June 2015.

[KRE 04] KREUTER M.W., MCCLURE S.M., "The role of culture in health communication", *Annual Review of Public Health*, vol. 25, no. 1, pp. 439–455, doi: 10.1146/annurev.publhealth.25.101802.123000, 2004.

[MIC 13] MICHIE S. *et al.*, "The behavior change technique taxonomy (v1) of 93 hierarchically clustered techniques: building an international consensus for the reporting of behavior change interventions", *Annals of Behavioral Medicine*, vol. 46, no. 1, pp. 81–95, doi: 10.1007/s12160-013-9486-6, 2013.

[MIS 16] MISPELBLOM BEYER F., *Encadrer les parcours de soins : vers des alliances thérapeutiques élargies ?*, Dunod, Paris, 2016.

[MON 12] MONDOUX A., "À propos du social dans les médias sociaux", *Terminal Technologie de l'information, culture & société*, vol. 111, pp. 69–79, 2012.

[MUN 17] MUNSON S.A., "Rethinking assumptions in the design of health and wellness tracking tools", *Interactions*, vol. 25, no. 1, pp. 62–65, doi: https://doi.org/10.1145/3168738, 2017.

[NIC 15] NICHOLAS J., LEDWITH A., BESSANT J., "Selecting early-stage ideas for radical innovation: tools and structures", *Research-technology Management*, vol. 58, no. 4, pp. 36–44, 2015.

[OCO 06] O'CONNOR G.C., DEMARTINO R., "Organizing for radical innovation: an exploratory study of the structural aspects of RI management systems in large established firms", *Journal of Product Innovation Management*, vol. 23, no. 6, pp. 475–497, 2006.

[OCO 08] O'CONNOR G.C., "Major innovation as a dynamic capability: a systems approach", *Journal of Product Innovation Management*, vol. 25, no. 4, pp. 313–330, 2008.

[OMS 46] OMS, "Preface to the Constitution de l'Organisation mondiale de la Santé", New York, no. 2, p. 100, 19 June–22 July, 1946.

[OST 05] OSTERBERG L., BLASCHKE T., "Adherence to medication", *The New England Journal of Medicine*, vol. 353, pp. 487–497, 2005.

[OST 10] OSTERWALDER A., PIGNEUR Y., *Business Model Generation: a Handbook for Visionaries, Game Changers, and Challengers*, John Wiley & Sons, New York, 2010.

[PAS 17] PASCAL P., CAPGRAS J.-B., CAZENEUVE H. *et al.*, Carnet Som'respir – projet PASCALINE, Assessment report, IFROSS Valorisation, December 2017.

[PHA 13] PHARABOND A.-S., NIKOLSKI V., GRANJON F., "La mise en chiffres de soi : une approche compréhensive des mesures personnelles", *Réseaux*, vol. 177, no. 1, pp. 97–129, 2013.

[PHA 16] PHAM Q., WILJER D., CAFAZZO J.A., "Beyond the randomized controlled trial: a review of alternatives in mhealth clinical trial methods", *JMIR mHealth and uHealth*, vol. 4, no. 3, p. 107, 2016.

[PIC 17a] PICARD R., *La co-conception en Living Labs santé et autonomie 1*, ISTE Editions, London, 2017.

[PIC 17b] PICARD R., *La co-conception en Living Labs santé et autonomie 2 – Témoignages de terrain*, ISTE Editions, London, 2017.

[PIN 17] PINTO G.L., DELL'ERA C., VERGANTI R. *et al.*, "Innovation strategies in retail services: solutions, experiences and meanings", *European Journal of Innovation Management*, vol. 20, no. 2, pp. 190–209, 2017.

[POM 15] POMEY M.-P. *et al.*, "Patients as partners: a qualitative study of patients' engagement in their health care", *PLoS ONE*, vol. 10, no. 4, p. e0122499, doi: 10.1371/journal.pone.0122499, 2015.

[RIF 15] RIFKIN J., *Zero Marginal Cost Society*, Palgrave Macmillan, 2015.

[SAO 15] SAOUT C., Cap santé, Report, available at: http://solidarites-sante.gouv.fr/IMG/pdf/20_07_15_-_RAPPORT_-_M-_Saout.pdf, 2015.

[SCH 18] SCHROEDER J., CHUNG C.F., EPSTEIN D.A. *et al.*, "An examining self-tracking by people with migraine: goals, needs, and opportunities in a chronic health condition", *Proceedings of the ACM Conference on Designing Interactive Systems (DIS)*, 2018.

[SÉN 18] SÉNAT, Rapport sur l'activité du Conseil d'Etat en 1991, available at: https://www.senat.fr/questions/base/1992/qSEQ920621702.html, 2018.

[SHO 06] SHORTLIFFE E.H., CIMINO J.J., *Biomedical Informatics: Computer Applications in Health Care and Biomedicine*, Springer, New York, 2006.

[SIL 05] SILVERSTONE R. (ed.), *Media, Technology and Everyday Life*, Ashgate, Aldershot, 2005.

[SLA 14] SLATER S.F., MOHR J.J., SENGUPTA S., "Radical product innovation capability: literature review, synthesis, and illustrative research propositions", *Journal of Product Innovation Management*, vol. 31, no. 3, pp. 552–566, 2014.

[SOL 15] SOLDHDJU K., *L'épreuve du savoir*, DingDingDong Editions, 2015.

[STO 94] STONE A.A., SHIFFMAN S., "Ecological momentary assessment (EMA) in behavioral medicine", *Annals of Behavioral Medicine*, vol. 16, pp. 199–202, 1994.

[TEE 10] TEECE D.J., "Business models, business strategy and innovation", *Long Range Planning*, vol. 43, nos 2–3, pp. 172–194, 2010.

[THI 15] THIEVENAZ J., TOURETTE-TURGIS C. (eds), *Penser l'expérience du soin et de la maladie : Une approche par l'activité*, De Boeck, Louvain-la-Neuve, 2015.

[TUR 11] TURKLE S., *Alone Together: Why We Expect More from Technology and Less from Each Other*, Basic Books Inc., New York, 2011.

[WYA 91] WYATT J.C., SPIEGELHALTER D.J., "Field trials of medical decision-aids: potential problems and solutions", *Proceedings of the Annual Symposium on Computer Applications in Medical Care*, pp. 3–7, 1991.

[VAN 15] VAN HOUTUM L. *et al.*, "Do everyday problems of people with chronic illness interfere with their disease management?", *BMC Public Health*, vol. 15, p. 1000, 2015.

[VER 08] VERGANTI R., "Design, meanings, and radical innovation: a metamodel and a research agenda", *Journal of Product Innovation Management*, vol. 25, no. 5, pp. 436–456, 2008.

[VER 11] VERGANTI R., "Radical design and technology epiphanies: a new focus for research on design management", *Journal of Product Innovation Management*, vol. 28, no. 3, pp. 384–388, 2011.

[VIL 18] VILLANI C., Donner un sens à l'intelligence artificielle : pour une stratégie nationale et européenne, Report, available at: https://www.economie.gouv.fr/files/files/PDF/.../Rapport_synthese_France_IA_.pdf, 2018.

[VIS 07] VISWANATH K., KREUTER M.W., "Health disparities, communication inequalities, and eHealth", *American Journal of Preventive Medicine*, vol. 32, no. 5, pp. S131–S133, doi: 10.1016/j.amepre.2007.02.012, 2007.

[WAR 17] WARE P. *et al.*, "Using ehealth technologies: interests, preferences, and concern of older adults", *Interactive Journal of Medical Research*, vol. 6, no. 1, p. e3, 2017.

[WES 14] WEST J., SALTER A., VANHAVERBEKE W. *et al.*, "Open innovation: the next decade", *Research Policy*, vol. 43, no. 5, pp. 805–811, 2014.

[WU 10] WU A.W., SNYDER C., CLANCY C.M. *et al.*, "Adding the patient perspective to comparative effectiveness research", *Health Affairs (Millwood)*, vol. 29, no. 10, pp. 1863–1871, 2010.

[WYA 91] WYATT J.C., SPIEGELHALTER D.J., "Field trials of medical decision-aids: potential problems and solutions", *Proceedings of the Annual Symposium on Computer Applications Medical Care*, pp. 3–7, 1991.

[ZIE 13] ZIEMER J., ERVIN J., LANG J., "Exploring mission concepts with the JPL innovation foundry A-team", *AIAA Space*, 2013.

List of Authors

Denis ABRAHAM
Institut Mines Télécom (IMT)
Paris
France

Sylvie ARNAVIELHE
Kyomed INNOV
Montpellier
France

Olivier AROMATARIO
EHESP
and
CRAPE – Arènes
Rennes
France

Mathias BEJEAN
Paris-Est Créteil Val-de-Marne
University
Paris
France

Anne-Marie BENOIT
UMR-Pacte, CNRS
Grenoble
France

Pierre BERTRAND
Laboratoire de Mathématique
University of Clermont Auvergne
Clermont-Ferrand
France

Marie-Noëlle BILLEBOT
ARS Nouvelle Aquitaine
Bordeaux
France

Nathalie BLOT
Société SESIN
Marseille
France

Karima BOURQUARD
IN-SYSTEM
and
IHE-Europe
Paris
France

Hugues BROUARD
Stiplastics Healthcaring
Royan
France

Linda CAMBON
ISPED
University of Bordeaux
France

Marie-Ange COTTERET
METRODIFF
Paris
France

Perrine COURTOIS
ELIA Consulting
Paris
France

Virginie DELAY
Stiplastics Healthcaring
Royan
France

Frédéric DURAND-SALMON
Bepatient
Paris
France

Guy FAGHERAZZI
Institut Gustave Roussy
Paris
France

Jean-Baptiste FAURE
Association ENCAPA
and
Fondation Partenariale i2ml
Toulouse
France

Matthieu FAURE
Fondation Partenariale i2ml
and
Institut Méditerranéen des Métiers de la Longévité
Nîmes
France

Bastien FRAUDET
ISAR
Pôle Saint Hélier
Rennes
France

Thierry GATINEAU
Laboratoire d'Innovation Technologique
Direction Data & Innovation
Harmonie Mutuelle
Saint Pierre Des Corps
France

Yves GRILLET
Fédération Française de Pneumologie
Paris
France

Caroline GUILLOT
Diabète Lab
Paris
France

Gaël GUILLOUX
L'École de design
Nantes Atlantique
France

Daniel ISRAËL
Société SIVAN INNOVATION
Jerusalem
Israel

Myriam LE GOFF-PRONOST
Laboratoire LEGO
IMT Atlantique
Brest
France

Loïc LE TALLEC
Bepatient
Paris
France

Myriam LEWKOWICZ
Institut Charles Delaunay, CNRS
Université de Technologie de Troyes
France

Henri NOAT
Société SESIN
Marseille
France

Norbert NOURY
INSA Lyon
France

Robert PICARD
Forum LLSA
and
Ministère de l'Économie
Paris
France

Hervé PINGAUD
Connected Health Lab
Ecole d'Ingénieurs en Informatique et Systèmes d'Information pour la Santé
and
INU Champollion
Castres
France

Patrick VISIER
ByBeflow
Paris
France

Index

S, T